# For Bradley M. Ditzel

Who has made his father so proud of his only son

# SELECTED BIBLIOGRAPHY OF BOOKS BY PAUL DITZEL

A Century of Service — The History of the Los Angeles Fire Department

Emergency Ambulance

Fireboats

Fire Alarm! The Story of a Fire Department

Fire Engines, Firefighters

How They Built Our National Monuments

Railroad Yard

The Complete Book of Fire Engines

The Day Bombay Blew Up

The Kartini Affair

True Blooded Yankee

Library of Congress Cataloging in Publication Data
Ditzel, Paul
Fireboats
Library of Congress Catalog Number: 89-080107
ISBN 0-925165-01-0

Published by Fire Buff House Division of Conway
Enterprises, Inc.
P.O. Box 711, New Albany, Indiana 47150
© Paul Ditzel 1989
First printing September 1989
Second printing November 1989

Printed in the United States of America

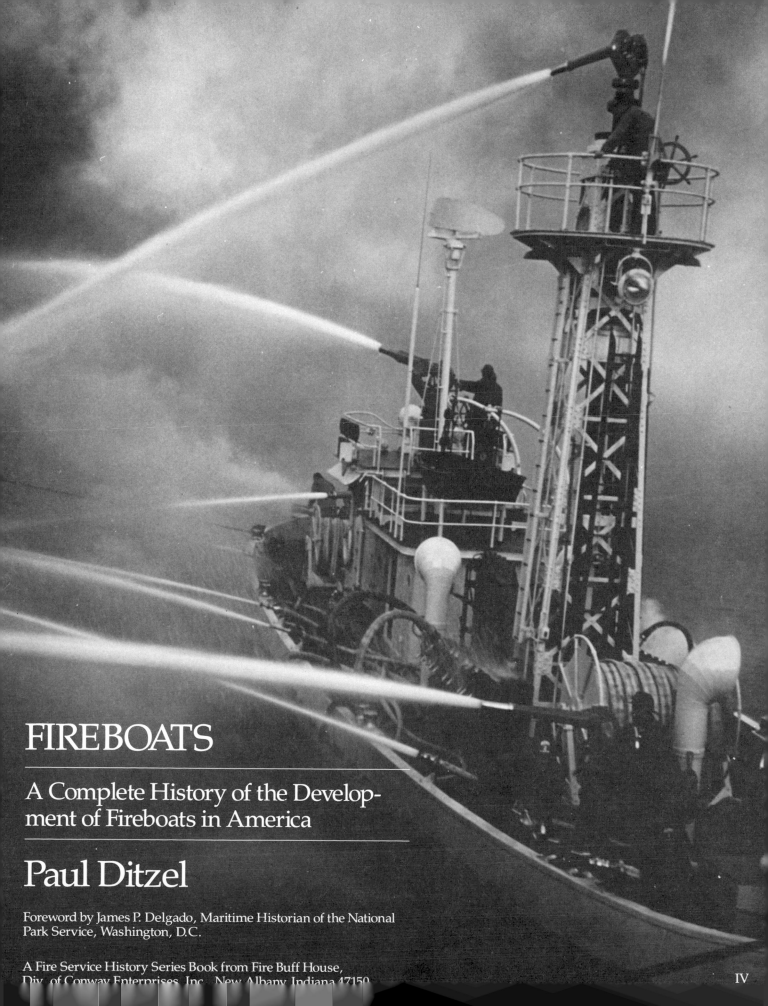

# FIREBOATS

## A Complete History of the Development of Fireboats in America

# Paul Ditzel

Foreword by James P. Delgado, Maritime Historian of the National Park Service, Washington, D.C.

A Fire Service History Series Book from Fire Buff House, Div. of Conway Enterprises, Inc., New Albany, Indiana 47150

# CONTENTS

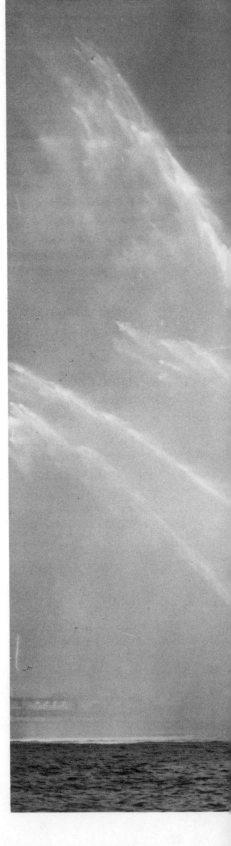

Foreword      VII

1. "Dispatch All Boats!"      2

2. Slings And Arrows      9

3. The Fireboats That Saved Chicago      24

4. The Anatomy Of A Fireboat      38

5. The Day The **Grattan** Blew Up      56

6. "We're At War!      74

7. Blanket Of Flames      89

8. Gallant Ship      108

Appendix — Log of American Fireboats      IX

Acknowledgements      XIII

Bibliography      XIV

Index      XXVIII

# FOREWORD

Some 184 fireboats were built in the United States between 1866 and 1989. As the date of the first boat's construction indicates, fireboats were the product of the Industrial Revolution, even though the concept of using vessels to fight fires on other vessels and along a port's waterfront dates to mid-18th century London. In the United States, pumps and hand-engines were placed on "floats" or small boats by New York volunteer firefighters as early as 1809. The 19th century development of large volume steam powered pumps provided sufficient pressure for effective firefighting. The first use of a floating steam pump to fight fires was aboard an unpowered London barge in 1852 that drew from an unlimited source, the Thames. Harbor tugs and towboats, the most common steam powered craft in any harbor, were the first fire fighting vessels in the United States. Very few vessels were designed as fireboats; rather, many tugs were fitted with pumps and monitors for auxiliary fireboat use. New York's first fireboat, for example, was a tugboat under contract to the port for firefighting.

The need for maximum capability to combat serious waterfront blazes on wooden ships and the wooden waterfronts of the late 19th and early 20th century compelled many fire departments in port cities to design and construct their own full-time fireboats. Naval architect Charles West, speaking to his colleagues in 1908, noted that the "comparatively temporary nature of American building construction" had led to the rapid development of fireboats in the United States. In 1896, naval architect H. De B. Parsons, speaking before the Society of Naval Architects and Marine Engineers, stated that "fireboats are of such importance to all marine cities, that they are properly regarded as a permanent and indispensable feature of their fire equipments."

Fireboats were built and employed on the Atlantic Seaboard, on the Gulf, Great Lakes, Pacific Coast, and on occasion on the inland rivers. Throughout the 20th century, an average of 33 American port cities had fireboats. The great port of New York has had the greatest number of fireboats, and continues to possess the nation's largest fleet today, while other ports, such as New Orleans, Philadelphia, Cleveland, Chicago, Buffalo, Seattle, Los Angeles, Portland, Oregon, and Baltimore have built several boats. In 1986, the Los Angeles Fire Department conducted a nationwide fireboat survey. A total of 27 cities in the United States that responded to the survey had 65 boats in service. Two cities, Tacoma and Seattle, Washington, were preserving laid up historic fireboats. Of the remaining vessels, only ten were fifty years old or older; most other fireboats date from the 1960s or later.

Of these ten fireboats 50 years old or older, only eight remain in service in 1989 — two New York boats, **John J. Harvey** (1931) and **Fire Fighter** (1938), New Orleans' **Del-**uge (1922), Los Angeles' **Ralph J. Scott** (1925), Portland, Oregon's **David Campbell** (1925), Seattle's **Alki** (1927), Mobile's **Ramona Doyle** (1939) and Buffalo's **Edward M. Cotter** (1900). Each of these vessels is historic and as much a part of the nation's maritime history and culture as the great squareriggers, river steamers, battleships, and tall-masted schooners that once plied our waters and which are today preserved and displayed at maritime museums around the country. Yet fireboats for the most part have been ignored in the recognition of the nautical past, relegated to the realm of fire history and the enjoyment of fire buffs who delight in the restoration of pumpers and engines of years past. Fireboats are appropriately a part of that history, but they also speak to the working waterfronts and the work-a-day craft that kept maritime trade, commerce, and naval defense active and healthy.

Tacoma, Washington, has moved its historic **Fireboat Number 1** ashore. Now displayed in a concrete basin, the fireboat is the only museum fireboat in the United States. Listed in the National Register of Historic Places — the first American fireboat so honored — **Fireboat Number 1** attracted attention of maritime preservationists and historians to the saga of the American fireboat. If plans are successful, a nearby port's fireboat will also be preserved as a museum display. **Duwamish,** built in 1909 for Seattle, is now laid up and awaiting a decision on her future. Seattle is home to a fleet of historic ships, including the lightship **Number 83, "Relief,"** the tug **Arthur Foss,** three-masted schooner **Wawona,** ferry **San Mateo,** and the schooner **Adventuress.** Hopefully a place can be found for another addition in the form of the city's famous fireboat. The other historic fireboats remain in operation, retooled with new engines and occasionally with new pumps and equipment.

To honor the contributions made to American maritime, naval, and firefighting history by these fireboats, the National Maritime Initiative of the National Park Service recently studied them as part of a special "Maritime Heritage of the United States" theme study done as part of the National Historic Landmarks survey. National Historic Landmarks are the most significant of the nation's recognized historic structures, buildings, sites, and objects. Seven fireboats were studied. Three represent the second generation of American fireboats; large steel-hulled, powerful pumpers as represented by **Edward M. Cotter,** formerly **William S. Grattan; Duwamish,** and **Deluge.** The significance of these boats as excellent examples of the type is enhanced by the national significance of the ports they served. Two gasoline-powered third generation fireboats were studied: **Fireboat Number 1,** which is the only boat to retain all of its original equipment, notably the gasoline engines, and **Ralph J. Scott,** formerly **L.A. City Number 2,** chosen as a representative of the type and for the importance of the port of Los Angeles and two of the nation's worst tanker fires which the boat fought.

Only one vessel, New York's **Fire Fighter,** survives that was designed and constructed as a fourth generation diesel-electric fireboat. The nation's best known fireboat, **Fire Fighter** represents a long and celebrated career capped with awards, a nationally significant port, and the culmination of classic fireboat design. One World War II fireboat was also studied. **City of Oakland,** formerly **Hoga, YT-146,** was included because of its noteworthy firefighting role at Pearl Harbor during the Japanese attack of December 7, 1941. The only known surviving Navy vessel afloat from the "Day of Infamy," **Hoga** saved men in the water, assisted three ships in distress, and fought fires for 72 hours on the **USS Nevada, Tennessee, Maryland,** and **Arizona.**

The seven fireboats were found to be nationally significant by the National Park System Advisory Board, a body that reviews all National Historic Landmarks. The Advisory Board and the National Park Service have recommended to the Secretary of the Interior that he designate all but one (the City of Buffalo objected to designating **Edward M. Cotter)** of the fireboats as National Historic Landmarks, helping insure their preservation and recognizing their unique contributions to American history.

The pending National Historic Landmark status of the fireboats recognizes the achievements of the boats and of their crews, among them men who lost their lives in the line of duty. As the son of a firefighter, I am pleased that the recognition of the fireboats extends to the hardworking crews who maintain and operate the fireboats. I met them in the various ports as I traveled across the country doing the study, riding aboard the **Ralph J. Scott, City of Oakland, Deluge,** and **Fire Fighter,** and enjoying traditional firehouse hospitality everywhere I went. It was an extra treat while in Los Angeles to meet Bill Dahlquist, one of the **Scott's** pilots and author of the Los Angeles Fire Department national fireboat survey, and Paul Ditzel. Paul and Bill were invaluable allies in the landmark studies of the fireboats. The reasons that the seven fireboats were selected for landmark study unfold in the pages that follow. But this book does more than document the facts and figures — it makes the boats live, and the heroic, difficult, dangerous, and often tedious duties of the firefighting mariners fill the pages. Each of the 186 fireboats has its own special and unique history. This is for them and the firefighters who race to burning ships or blazing piers with the spray flying over the decks and monitors blasting.

James P. Delgado
Maritime Historian of the National Park Service
Washington, D.C., May 15, 1989

*Los Angeles' fireboat, the **Ralph J. Scott,** demonstrates its powerful water wallop of nearly 20,000 gallons-per-minute. From the Collection of Paul Ditzel.*

# CHAPTER ONE
## "DISPATCH ALL BOATS!"

The pre-dawn fire, seen by many but reported by nobody, smoldered for hours under a wharf at Berth 73 in the heavily-industrialized Port of Los Angeles. Commercial fishermen and others who noticed something white oozing from the base of a high voltage light post on the wharf dismissed what they saw as trivial, even though they also heard crackling and popping. No point in bothering firefighters, they agreed. High tide was due. That would surely squelch whatever the problem was under the wharf.

Smoke worsened as the fire gestated along the spiderweb of wooden joists, stringers and pilings that supported the asphalt-covered timber deckwork. These underpinnings were coated with wood-preserving creosote, a highly-flammable coal tar derivative and a known carcinogen. Gnawing with increasing speed and fury along the wharf's underbelly, what began as a relatively insignificant electrical short circuit suddenly burst into a full-blown fire with disastrous potential.

Flames stabbing from under the wharf were crowned with thick clouds of creosote-impregnated smoke. Heat radiated out toward commercial fishing boats that were moored alongside the 1700-foot-long wharf. Still nobody turned in an alarm. At least two fishermen were trying to reach the fire with a small hose when, hundreds of yards away, a security guard saw smoke veiling the glare and called the Los Angeles Fire Department.

At 4:19 that Sunday morning, January 31, 1988, the alarm was broadcast to the nearest fire companies. In additional to five engines and two aerial ladder trucks, the alarm called out two fireboats including *Boat 2*, the *Ralph J. Scott*, one of the world's most powerful. Over half-a-century old and named after a former fire chief, *Boat 2* packs a wallop of more than one million gallons of water an hour through 13 guns.

Hurrying from their Berth 85 firehouse along the Main Channel, the *Scott's* eight crewmen boarded the 99-foot-long craft and prepared to cast off. Pilot Bill Dahlquist, standing at the control console in the pilothouse, switched on the three diesel propulsion engines. Below deck, the aft-mounted, side-by-side engines began to warm up.

Also below deck and inside the glass-enclosed Quiet Room just forward of the propulsion powerplants, Engineers Jim Choner and John Rassmussen took up their stations. Arrayed in front of them on the bulkhead-mounted console were pumping and propulsion throttles, gauges and a rainbow of walnut-shaped indicator lights. Choner and Rassmussen began to idle the four diesel pumping engines located in pairs toward the bow. Altogether, the seven engines, even at idle, raised the noise level outside the Quiet Room to a thunderously high decibel reading.

Berth 73 was a quick run for the nearest land-based fire company, Task Force 48, consisting of two engines and an aerial ladder truck. Pulling onto the wharf and seeing the scope of the rapidly-spreading fire and sensing the difficulty of reaching it with land-based companies, Task Force Commander Gerald Ramaekers radioed the LAFD alarm office: "Dispatch all boats!...This is a major emergency!"

*Pilot Bill Dahlquist of the LAFD's **Fireboat No. 2** at the control console in the pilothouse. The boat, one of the world's most powerful, has a pumping capacity rated at 18,655 gallons-per-minute. LAFD*

And thus was set in motion the seldom seen response of all five fireboats in the LAFD fleet. He also called for more land companies.

Ramakers' broadcast was heard in *Boat 2's* pilothouse and in the Quiet Room. The fireboat firefighters knew that a long, hot and smoky fire lay ahead of them. Dahlquist eased ahead on *Boat 2's* throttles. The *Scott* quickly gathered speed as Dahlquist steered a wide, sweeping arc to starboard. He set the boat's course along mid-channel to avoid creating wake damage to small boats moored alongside and near the famous tourist attraction, Ports O'Call Village, with its restaurants and small shops.

Glancing off the starboard bow, Captain Don Hibbard saw a column of smoke masking a glare. Judging from the size of the loomup, the veteran waterfront firefighter expected this fire would, as the history of wharf fires has shown many times over, take many hours to control. Concealed in the thick creosote smoke, moreover, were the omnipresent dangers of fighting wharf fires which historically cause many injuries and sometimes deaths. At the

*Glass-enclosed for soundproofing, the LAFD's Quiet Room, below the deck of* **Fireboat No. 2**, *contains the bulkhead-mounted console with pumping and* *propulsion throttles (foreground) gauges and indicator lights. Note annunciator (lower left). From the Collection of Bill Dahlquist*

very least, irritants in the smoke, even with protective breathing apparatus, could cause skin burns and respiratory problems for firefighters.

Dead ahead, as *Boat 2* neared the entrance to the Southern Pacific Slip leading to Berth 73, Dahlquist saw the rotating blue light and silhouette of fast-approaching *Fireboat 1*, with its two crewmen and Supervising Mate Fred Stoddard at the helm. The speedboat-sized craft can pump 750 gallons a minute, but its primary purpose is as a launching platform for SCUBA (self-contained underwater breathing apparatus) divers. While the boat skimmed along the Main Channel its two firefighters were slipping into their wet suits preparatory to floating a highly hazardous attack directly underneath the wharf.

Entering the slip, fireboat firefighters saw many immediate and potential problems. Both sides of the narrow slip were cluttered with dozens of fishing vessels, mostly old mostly big and mostly wooden. Snuggling together by as many as three abreast, they were moored in an almost solid bow-to-stern line along the wharf. The six nearest the fire were in imminent danger of bursting into flames. The heat raised paint blisters on the wooden hulls and wheelhouses of the fishing boats. Somehow these—and other boats in the slip—would have to be moved before the *Scott* could effectively bring its underwharf nozzles to bear.

The fire was spreading several hundred feet north and south along Berth 73. Lapping flames threatened igloo-shaped piles of fishing nets on top of the wharf. In the

*Mate Mike Corcoran of LAFD **Fireboat No. 2** slips into breathing apparatus preparatory to operating bow turret gun. Bulwark nozzles (left and right) sweep wharf flames while underwharf nozzle (lower right) makes a direct attack on fire underneath the wharf. Mike Meadows*

northern path of the fire stood the Fisherman's Cooperative Association Building. Nudged by a slight breeze, flames were traveling faster in a southerly direction toward the end of the wharf where an oil storage tank farm jutted into the Main Channel.

About the only favorable elements in the worsening scenario were the high tide and the wood barrier skirts at several hundred feet intervals along the underside of the wharf. Extending down towards the water, the skirts would at least slow the snaking heat and flames. The high tide, moreover, narrowed the fire-feeding oxygen space between the wharf and the water.

Battalion 6 Commander Bill Burmester, calling for additional apparatus, created a North and a South Division of land companies to mount an assault from each end of the wharf. Captain Hibbard hurried to the stern of *Boat 2* where, as Marine Division Commander, he would have the best overall view while directing all fireboat and SCUBA operations with his handie-talkie radio.

As flames lapped out from under the wharf, Hibbard doubted that the *Anna Maria* and at least four other large fishing boats could be saved. The rapidly worsening situation allowed no time to await the arrival of Coast Guard help in moving the vessels. Radiated heat barred anybody from approaching them to release mooring lines.

Bringing Boat 2 around to starboard until the bow pointed at the fire, Dahlquist saw a 30-foot opening between the bow of the *Anna Maria* and the stern of another big boat. Hibbard decided to attack there while daring to hope that streams from *Boats 2* and *1* would take some of the energy out of the fire before the fishing boats burst into flames. It was not to be. While Dahlquist eased the *Scott* slowly forward, the *Anna Maria* and the boat ahead of it ignited. Almost immediately, heat touched off large piles of fishing nets stacked on the dock.

More bad news was imminent as *Boat 2* plowed into the smoke. Fireboat firefighters felt the combined heat from the blazing wharf dead ahead and that from the burning fishing boats a few feet off their port and starboard sides. Mate Mike Corcoran stood ready at the bow turret gun while Firefighters Glenn Thorson, Frank Vidovich and "Buzz" Smith crouched at the bulwark-mounted nozzles.

Dahlquist alerted on-deck firefighters that he was about to call for pumping to begin and gave four blasts on the boat's air horn; the warning signal to land companies that heavy stream operation was going to start. Dahlquist rang the pilothouse annunicator which sounded a bell in the Quiet Room. Choner and Rassmussen saw their annunciator's red arrow move to "START". The engineers pushed the pumps' throttles forward. Dahlquist rang three more times and the arrow swung around to 50 pounds pressure-per-square-inch (psi), then to 100 and finally to 150 as water surged through the mains.

A 3000-gallon-per-minute stream gushed from the bow turret as Corcoran directed the solid stream which knocked down fires among the stacks of fishing nets. The bulwark nozzles began spurting 750-gallons-per-minute at 125-150 psi. Five yellow lights and gauges on the Quiet Room control panel indicated the bow turret, three bulwark nozzles and an underwharf nozzle which was remotely controlled by Dahlquist in the pilothouse, were delivering battering ram streams. These salvos splattered against pilings stringers, and joists. Spray ricocheting into the firefighters' faces smeared the pilothouse windows.

Still more yellow lights blinked on in the Quiet Room as Dahlquist activated the bow port and starboard underwater thruster nozzles. Alternately using these jet streams to hold the boat in position, Dahlquist rotated the boat from side to side to bring more nozzles to bear in a cascading water sweep. With the bow buried in smoke, it was impossible to tell exactly how close *Boat 2* was to the wharf. The fender on the bow protected the *Scott* as it bumped into the wharf before Dahlquist eased back.

The awesome roar of the fire streams slamming against the pilings and the churning of the pumping and propulsion engines precluded voice communications between the pilothouse and on-deck firefighters who guided Dahlquist with hand signals.

In the Quiet Room, meanwhile, the engines' heat made the cubicle sauna-like as the boat's ventilation system sucked in wisps of eye-smarting creosote smoke to add to the engineers' discomfort. Contact with the pilothouse was by bell and a voice-magnified phone. Though sound-proofed with glass and other protections, the din of the pumping and propulsion engines created a nerve-wracking invasion of the otherwise Quiet Room. Outside the room, the high-pitched whine required ear protectors as visual checks of the engines' performance were regularly made.

The three other fireboats in the LAFD fleet were arriving and calling Hibbard for orders. Already in battle, *Boat 1* was directing its 750-gallon-per-minute bow monitor stream onto the blazing *Anna Maria*. The fireboat's sister vessels, *Boats 3* and *5* were ordered to prepare for deployment of SCUBA divers to set up water curtains at the north and south flanks of the fire. *Boat 4*, the 9000-gallon-per-minute *Bethel F. Gifford*, was to assist *Boat 2* in the frontal assault.

*The 9000-gallon per-minute **Bethel F. Gifford**, also known as **Boat No. 4** of the LAFD, operates two rail standee nozzles and one of its hull-mounted underwharf nozzles during Berth 73 wharf fire. Mike Meadows*

The LAFD's Fireboat No. 5 (upper left) was ordered to save its sister craft, Fireboat No. 1 (lower right) when Boat No. 1 was rammed during a wharf fire. Boat No. 1 reported it was fast taking on water and sinking. From the collection of Bill Dahlquist

Mate Mike Svorinich of **Boat No. 4** guides the **Gifford's** pilot as the big boat's bow is buried in thick clouds of creosote-impregnated smoke gushing out from under the blazing wharf. Mike Meadows

Suddenly came an alarming report. Stoddard radioed Marine Division Commander Hibbard that *Boat 1* had been rammed by a fishing vessel. *Boat 1* was taking on water fast and sinking. Hibbard ordered *Boat 5* to go to the aid of her stricken sister craft. While Hibbard anxiously awaited a further damage assessment, the *Scott's* guns continued to pound the fire. *Boat 1's* radio message was encouraging. The bilge was flooding, not from the ramming, but because a pump hose had snapped. *Boat 5* pumped out the bilge while repairs were made. Both craft soon were again ready for battle.

Perhaps the best illustration of LAFD underwharf firefighting tactics is to compare them to a naval assault against an entrenched enemy. Just as the heavy guns of battleships soften the beachhead preparatory to an invasion, so do the LAFD's two largest boats, the *Scott* and the *Gifford*, batter underwharf fires. SCUBA firefighters from the three smaller craft, meanwhile, carry the attack to the vulnerable flanks of the fire. Later, when the big boats have done their job, the SCUBA divers will move in to hit stubborn pockets of fire that master streams did not reach because the underwharf's gridwork deflected them.

And so it was before dawn that three teams of SCUBA divers—each with four polyurethane floats with nozzles and interconnected by hoselines supplied by the small boats—dumped their gear overboard and leaped into the water. Much like surfboarders, the SCUBA divers clung to their floats while using the backward thrust of water pressure to propel themselves into the darkness under the wharf.

Each two-man team—in line with skin diving's Buddy System for safety—lashed their large floats to pilings at the north and south flanks of the fire. Cone-shaped sprays

Flames and creosote-laden smoke lap from under blazing wharf as LAFD firefighters mount an attack from the top of the wharf. Mike Meadows

Tony Zar (left) and John McDaniel jackhammer holes in Berth 73 deckwork so other firefighters can access underwharf blaze with Bresnan distributors and other nozzles. Mike Meadows

*LAFD SCUBA firefighter uses back water pressure in hose to propel himself under wharf while using an attack float to knock down residual fire. Mike Meadows*

*Anna Maria* was being moved, flames reached stores of ammunition used by fishermen to shoot sharks preying upon their catch. Hundreds of rounds of exploding ammunition added to the pandemonium in the narrow slip as fishermen, swarming aboard their vessels, stampeded to safety in the Main Channel.

Dozens of fishing boats moving helter-skelter toward the channel created still more problems for fireboat firefighters. Many boat owners, panicking, skimmed dangerously close to the fireboats. Dahlquist constantly glanced aft to make certain the stern area was clear as he maneuvered the *Scott* to better focal points of attack along the burning wharf which was now clear of boats.

The Marine Division's battle proceeded in tandem with that of land-based North and South Divisions. Calls for more help resulted in a total LAFD response including 21 engines, eight aerial ladders, auxiliary equipment, including a rig to supply and refill breathing apparatus air bottles, and a helicopter for aerial reconnaissance.

Moving in from the north and south topsides of the wharf, firefighters opened pizza pan-sized metal lids which provided underwharf access. Attaching Bresnan nozzles to their 2½-inch hoselines, firefighters lowered them into the smoke and heat spewing from the openings. The nine ports on each whirling Bresnan directed upward and outward spray patterns of between 600 and 800 gallons-per-minute. Soon, engine companies would begin supplying hoselines of the SCUBA firefighters to free the small boats for better observation of the divers and other jobs consistent with the waterside attack.

Arriving from their Hollywood station, Heavy Utility 27's Apparatus Operator Tony Zar and his partner, Firefighter John McDaniel, began jackhammering additional holes along the inches-thick wharf. Each hole enabled another Bresnan to begin spinning a fire-suppressing water web under the wharf.

Slowly and arduously, the combined teamwork of marine and land companies gained the upperhand. When no more flames were visible from the slip, SCUBA divers guiding small attack floats and using the back pressure in their hoselines, moved in from both flanks to kill residual fires. The fire was fully-controlled shortly after dawn—less than three hours after the first alarm.

This was a remarkable achievement. If history had repeated itself, a wharf fire of this magnitude and spreading among massive numbers of old wooden fishing boats would have taken perhaps 12 hours to control. Dozens of firefighters could have been expected to be injured and perhaps several killed. Property losses would have been catastrophic. Testifying to the successful outcome are at least three facts:

First, there were no significant injuries. Only six of around 100 fishing vessels were damaged. Most of them could be repaired. And thirdly: from the time firefighters found flames moving along a several-hundred-foot-long front until containment, the fire spread only an additional 40-90 feet.

That this traditionally worst possible scenario resulted in an exemplary outcome demonstrates how far waterfront firefighting tactics have progressed since those first hand-operated fireboats went into service more than a century ago.

from each float-mounted nozzle overlapped as the SCUBA divers formed a water curtain past which the flames would not pass.

Glancing out *Boat 2's* port window, Dahlquist saw the *Anna Maria's* mooring line burn, separate and drop. Seizing the opportunity, Dahlquist brought *Boat 2* around to port and nosed the bow against the *Anna Maria's* hull, forcing the blazing fishing boat to swing clear of the wharf. On-deck firefighters, together with those on the wharf, played cooling streams onto the burning and smoking *Anna Maria*, which repeatedly bumped *Boat 2*.

Hibbard directed *Boat 5* to tow the *Anna Maria* to the opposite side of the slip where the remaining fire would be controlled with the aid of land-based companies. As the

# CHAPTER TWO
## SLINGS AND ARROWS

There was cause for professional pride as America's foremost fireboat designers met in New York, November 12 and 13, 1896, during the fourth annual meeting of the Society of Naval Architects and Marine Engineers. After years of rebuffs by city officials, the designers' concepts for powerful increased pumping capacity fireboats had not only been accepted but they were creating a new look along waterfronts as 25 of these boats went in service by that year. Fourteen more would come on line during the next decade.

"Firefighting in the marine cities is a new science since the modern fireboats came in," said William C. Cowles, who had designed or supervised construction of nearly half of these boats, including the much-publicized *New Yorker.* Chicago and Philadelphia each had four of the new breed of fireboats. New York had three. Boston, Brooklyn, Buffalo, Cleveland and Milwaukee had two apiece. Baltimore, Detroit and Seattle had one. Southern and Gulf ports were conspicuously absent from the list of fireboats and their specifications compiled by Naval Architect Harry deBerkeley Parsons, 34, as part of the paper he presented, "American Fire-Boats." The lack evidenced the south's slow recovery following the Civil War.

Largest of these boats was *The New Yorker,* 125½-feet-long with a 26-foot beam, 11½-foot draft, and displacing 470 tons. The steel-hulled *New Yorker* had two Scotch boilers, each with four furnaces and tubes purposely placed to maximize a high water level. To save weight *The New Yorker* normally operated with one boiler while the other remained empty and in reserve for extraordinary firefighting demands requiring more steam.

Not until the architects broached the subject of firepower did the capabilities of *The New Yorker* become awesome. Mounting a pair of vertical duplex (two-stage) pumps—two Clapp & Jones and two built by the La France Manufacturing Co.—the boat was rated at 11,530 gallons-per-minute (gpm) at 300 pounds pressure per-square-inch (psi). *The New Yorker* had plenty of reserve power to do over 16,000 gpm: the equivalent pumping capability of 14 land-based fire engines. By contrast, America's first steam fireboat, Boston's *William L. Flanders,* which went in service in 1873, put out 2500 gpm; hardly more than several land pumpers.

Parsons' compilation of the other boats ranged upwards from 75 feet with median pumping capacities around 5000 gpm. Fourteen had wood hulls; the rest were steel, including Buffalo's *George R. Potter* and *John M. Hutchinson.* Both also served to plow through ice as thick as 12 inches to open the port for Great Lakes shipping and to occasionally rescue fishermen—and sometimes their dogs—from ice floes.

Pumps were mostly supplied by some of the best-known steam fire engine builders, notably Clapp & Jones, Hudson, N. Y., the favored supplier; Amoskeag Manufacturing Works, Manchester, N.H.; La France Manufacturing Co. Elmira, N. Y.; Silsby Manufacturing Co. Seneca Falls, N. Y. and Thomas Manning Jr. & Co. Cleveland.

Like other harbor craft, fireboat power plants were outfitted with a variety of boiler types. The most common in Parsons' study was the Scotch fire tube boiler. This boiler consisted of a cylindrical fire tube boiler. This boiler consisted of a cylindrical tank partly filled with water through which the furnaces and tubes passed hot gasses to the smokestack or stacks. Steam was drawn from the top of the boiler to power the engines. Fireboats always kept one boiler at the ready to allow quick response to an alarm.

Simple expansion steam propulsion engines were favored in 17 of the 24 boats. In simple expansion systems, steam is used once in a cylinder, then is exhausted up the stack or returned to the boiler by a condenser. These engines offered the advantages of low initial cost and were easy to operate and maintain.

Compound engines in the other boats functioned when steam, expanding in one cylinder, was channelled (ported) into the intake of an adjacent cylinder where it expanded again. Double and triple compounding allowed steam to be used (expanded) more than once, thus yielding greater economy during prolonged operation. The architects noted that because nearly all fires were of around one hour's duration, operational economy did not offset higher purchase costs and increased complexity. Only *The New Yorker* had triple compounding among these early fireboats. That seldom-needed capability proved useful, shortly after the boat was commissioned, when *The New Yorker* pumped continuously for five days during the New York Central Railroad's grain elevators fire.

"She is a noble boat and has done great service," said Cowles. "The (fire insurance) underwriters will tell you that." Only six years old, *The New Yorker* had repaid its $105,000 cost in property saved and fires prevented from spreading."

Parsons recalled when "'The New Yorker' lay alongside of a burning ship in the East River some two or three years ago that was so hot that the land companies could not approach it. She stayed there until the ship was about to fall and got away just in time to save herself from being caught in the rigging. That is probably the narrowest escape 'The New Yorker' has had since her existence."

Ship-to-shore contact and among fireboats was by megaphones, signal flags, lights and, shortly after 1900, wireless telegraph. Orders from the pilothouse were communicated to the boiler room by speaking tubes and annunciators or pump telegraphs which were often chain-operated and pie-shaped. Each wedge of the pie indicated an order: start or stop pumps and pumping pressure increments beginning at 50 pounds and increasing to 250 or more. The pilothouse office called for the desired operational mode by turning a handle which not only moved an arrow to the marked wedge but simultaneously registered on the boiler room annunciator along with an attention-getting bell that could be heard over the raucous churning of the engines and pumps.

Coal invariably was stored in watertight compartments on the port and starboard of the boiler area. It was supplied to the boiler room deck by at least one chute and from there was hand-shoveled into the furnaces. The job was exhausting and there was little room for stokers to move around. Below deck temperatures could exceed 130 degrees Fahrenheit.

**The New Yorker,** with a pumping capacity exceeding 16,000 gpm, was the most powerful fireboat of her era and one of the first to be especially built for firefighting. Commissioned February 1, 1891, **The New Yorker** could produce as much water as 14 land-based fire engines. Among its features were swiveled nozzles (inset) and iron shields to protect firemen from heat as they operated the nozzles. From the Collection of Paul Ditzel

Boston's **William M. Flanders** was the first steam fireboat in the United States. In service January 1, 1893, the 75-foot iron-hulled fireboat had four Amoskeag pumps delivering 2500 gpm. The **Flanders** displaced 55 tons and cost $19,893.95. Inset photos are believed to be those of the first officers and crew. From the Collection of Bill Noonan

*Buffalo's **George R. Potter,** named after a fire commissioner, had a low profile typical of Great Lakes' fireboats to enable it to quickly pass under bridges. The **Potter,** built in 1896, had a steel hull, was 80-feet-long and had a 20-foot-beam. Displacing 130 tons, it had a pumping capacity of 3810 gpm. Buffalo Fire Historical Society*

*American Fire Engine Co., Seneca Falls, N.Y. (later American LaFrance) became a supplier of pumps to fireboats of New York, Chicago, Boston, Philadelphia and other cities. Illustration from 1902 company catalog shows two of the latest and largest units built by the company. Each pump delivered 2500 gpm at 200 psi, was self-contained, and stood 9-feet, 6 inches. From the Collection of George Anderson*

For 26 years *The New Yorker* was berthed at the southerly tip of Manhattan Island. This first Fire Department of New York "Super Fireboat" served until December 17, 1931, and was auctioned for $590. From the Collection of Paul Ditzel

Not only was ventilation virtually non-existent but dangerous situations lurked. Marine Engineer John D. McKean was fatally scalded, September 17, 1953, by a boiler room explosion aboard New York's *George B. McClellan*, a half-century-old boat nearing the end of its service. Later, a fireboat was named in McKean's memory. In 1988 it was still on the FDNY's Marine Division roster.

Cowles said an immense increase in boiler capacity was not only the secret of success of the new fireboats but the significant difference between them and tugboats which heretofore stood as the first line of waterfront protection along with steam fire engines on barges which were towed to fires.

"The old fireboats (before 1884) were simply tugs with the ordinary boiler capacity of a tug (and with a few water-shooting nozzles). In the new fireboats the idea was to get as large a boiler capacity as possible said Cowles. "In my designs (the boiler) capacity was doubled and even quadrupled without increasing the size of the hull."

Even at that, Cowles and his fellow naval architects agreed that fireboat boilers were still undersized because the pumps and propulsion engines could quickly exhaust steam supply in boats other than *The New Yorker*.

Cowles advised his colleagues that "You want to get the most firefighting capacity you can for the money and you do not want to put it all in your engines. Your boiler comes first and your pumps next. Your engines and hull are simply things to push and float your firefighting machine."

Of the 24 boats examined by the naval architects, only Brooklyn's *Seth Low*—designed by Cowles—a 99-foot-long craft rated at 3360 gpm and named after a former mayor of that city, had twin propulsion screws. It was agreed that twin screws enhanced maneuverability, but as Parsons pointed out, "The extra floor space weight and difficulty of management of the double machinery offsets all the advantages gained."

Based upon their cumulative experience, some of it gained by trial-and-error, the designers turned to speed and fireboat lengths. Fourteen statute miles per hour was tops they generally agreed, depending largely upon the size of the waterfront area requiring coverage. Greater coverage probably dictated a second boat, as did any fireboat over 100-feet long. The greater the length the more difficult it was to maneuver in slips and similar narrow areas commonly found along waterfronts.

"Fireboats to be most efficient for the purpose for which they were built, should be of a size that are quick to handle," said Parsons. "This consideration limits their length. And it appears from our experience in handling these boats that a capacity of 6000 gpm is sufficient. That capacity will make two sets of duplex pumps—3000 gallons to a set....At a large fire, a 6000-gallon capacity will always be required."

The discussion was starting to suggest that *The New Yorker*, for all its vaunted pumping capabilities, was a freak. Retrospectively, Cowles made an observation that would be repeated by generations of other fireboat designers: "My

Naval Architect Harry deBerkeley Parsons designed New York's **George B. McClellan,** named after a New York mayor. Featuring a pumping capacity of 7000 gpm, it was among the most powerful fireboats when commissioned, October 1, 1904. The steel-hulled boat was 117-feet-long, had a 25-foot beam and a draft 10- feet, 6-inches and had three monitors. After 51 years of service, a September 17, 1953, boiler explosion killed Marine Engineer John B. McKean. The **McClellan** was disposed of the following year. From the Collection of Steven Lang.

Brooklyn's first fireboat was the **Seth Low,** with a minimum pumping capacity of 3360 gpm via its twin Clapp & Jones steam-powered pumps. In service in 1886, the 99-foot-long **Seth Low** featured two propellers, a unique capability at that time when all other boats were built with one propeller. The fireboat had a wood hull, a beam of 23-feet and a 7¹/₂-foot draft. From the Collection of Herbert J. Eysser

## FLOATING STEAM FIRE ENGINES.

SHAND, MASON & Co.'s DOCK FLOATING STEAM ENGINE is available in crowded Docks and Harbours where a vessel could not be moved at high speed, and is much less costly than a screw or paddle boat. In this the ordinary propeller and engines are dispensed with, and the requisite fittings introduced so as to make the Fire Engine available as a motive power on the water-jet propulsion system.

The First Floating Steam Fire Engine for the London Fire Brigade was constructed by SHAND, MASON & Co. on this principle, when with a boat, large out of all proportion in comparison with its fire engine, a speed of 4 miles an hour was obtained. The large Floating Fire Engine of the Metropolitan (London) Fire Brigade, illustrated at page 34, is propelled in this manner at a rate of about 12 miles an hour, a centrifugal pump being used for the water-jets: but by the mere use of jets from the fire engine on SHAND, MASON & Co.'s system, a speed of from 4 to 6 miles an hour is obtained.

The boat is of iron, with an average length of about 40 ft., and breadth of beam of about 10½ ft.; the machinery is all enclosed, and there is ample space for fuel, tools, hose, and all necessary articles. The Engine is horizontal, and, with the Boiler, is of the same construction as the ordinary Steam Fire Engine. It draws from the water in which it floats by a fixed suction pipe, or from any other source by a flexible one. The prices include all the usual apparatus, with the exception of hose and suction pipes, the prices of which are at the foot of this page.

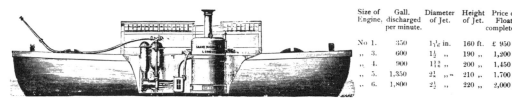

| Size of Engine. | Gall. discharged per minute. | Diameter of Jet. | Height of Jet. | Price of Float complete. |
|---|---|---|---|---|
| No 1. | 350 | 1¼⁄₁₆ in. | 160 ft. | £ 950 |
| ,, 3. | 600 | 1½ ,, | 190 ,, | 1,200 |
| ,, 4. | 900 | 1¾ ,, | 200 ,, | 1,450 |
| ,, 5. | 1,350 | 2¼ ,, | 210 ,, | 1,700 |
| ,, 6. | 1,800 | 2½ ,, | 220 ,, | 2,000 |

DOCK FLOATING STEAM FIRE ENGINE, AS USED BY THE LONDON AND ST. KATHERINE'S DOCK COMPANY, &c.

LAND STEAMER PLACED IN A BARGE AND USED BY THE METROPOLITAN (LONDON) FIRE BRIGADE AS A FLOATING STEAM FIRE ENGINE.

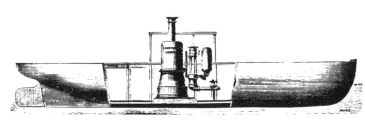

LAND STEAMER, WITHOUT WHEELS, AXLES, AND FRAMING, FIXED IN A BOAT; AS USED BY THE AUSTRIAN AND OTHER GOVERNMENTS.

All of SHAND, MASON & Co.'s LAND STEAM FIRE ENGINES can be prepared as above, ready for fixing into boats, with the necessary connections for fixed and flexible suction pipe, and with all the usual implements, at the under-mentioned prices—

| SIZES OF ENGINES. | No. 1. | No. 2. | No. 3. | No. 4. | No. 5. | No. 6. |
|---|---|---|---|---|---|---|
| Indicated Horse power . . . . . | 30 | 40 | 55 | 82 | 120 | 160 |
| Diameter of Jet when only one is used . | 1¼⁄₁₆ inch. | 1½ inch. | 1½ inch. | 1¾ inch. | 2¼ inches. | 2½ inches. |
| Height of Jet . . . . . . . . | 160 feet. | 170 feet. | 190 feet. | 200 feet. | 210 feet. | 220 feet. |
| Gallons delivered per minute . . . . | 350 | 450 | 600 | 900 | 1,350 | 1,800 |
| Price of Engines { London Brigade Vertical . . . . | £105  0  0 | £450  0  0 | — | — | — | — |
| Patent Single Horizontal . . . | 375  0  0 | — | — | — | — | — |
| Patent Equilibrium . . . . . | — | 540  0  0 | £630  0  0 | £810  0  0 | £1,000  0  0 | £1,260  0  0 |
| Double Horizontal . . . . . | — | — | — | — | — | — |
| Best India-rubber or Leather Suction Pipe in 10 ft. lengths, with couplings complete . . . . . . . . | 7 18 0 | 8 18 0 | 9 18 0 | 12 0 0 | 13 0 0 | 14 10 |
| Best copper-riveted leather Hose, in 40 ft. lengths, with couplings, hand-loops, and straps complete . . . . | 8 10 0 | 8 10 0 | 8 10 0 | 8 10 0 | 8 10 0 | 8 10 0 |
| Best woven Canvas Hose, prepared to prevent rot, in 100 ft. lengths, with leather strap and couplings complete { No. 1 quality | 8 10 0 | 8 10 0 | 8 10 0 | 8 10 0 | 8 10 0 | 8 10 0 |
| ,, 2 ,, | 7 5 0 | 7 5 0 | 7 5 0 | 7 5 0 | 7 5 0 | 7 5 0 |
| ,, 3 ,, | 6 0 0 | 6 0 0 | 6 0 0 | 6 0 0 | 6 0 0 | 6 0 0 |
| Best patent tanned Canvas Hose, lined with India-rubber, 100 ft. lengths, with couplings complete . . . . . | 11 0 0 | 11 0 0 | 11 0 0 | 11 0 0 | 11 0 0 | 11 0 0 |

*When London fire officers discovered that hand pumps and beer drinking did not mix, they ordered a steam fire engine mounted on a barge. This floating steam fire engine was built by England's famed fire apparatus maker, Shand, Mason & Co. The apparatus was propelled by water jets. From the Collection of Bill Dahlquist*

**FLOATING STEAM FIRE ENGINES** are a necessary acquisition in Ports, Docks, &c., where warehouses and stores of goods are in proximity to water. They are made self propelling, or are placed in a vessel to be moved about by steam tugs, but frequently the Engines alone, without the boilers, are fitted on board steam tugs and ferry boats, the steam from the propelling engine boiler rendering them immediately available in case of fire. This latter arrangement is also used in tank boats for supplying ships with fresh water, pumping from the sea to extinguish fires, and for removing water from the holds of ships.

The first Floating Steam Fire Engine dates from 1852, when SHAND & MASON applied steam power to one of the Manual Floating Fire Engines of the London Fire Brigade; and the result of this experiment was so satisfactory that a complete self-propelling Floating Steam Fire Engine was immediately ordered from the same firm. This is still (1883) the most powerful and efficient Steam Floating Fire Engine in London, after having been employed for twenty-five years at every river-side fire since its construction, many of them being at a considerable distance from the Thames.

The most complete self-propelling Steam Fire Engine in existence, was constructed by SHAND, MASON & CO., for Her Britannic Majesty's Government for use at Calcutta, and on account of the rapid current of the Hooghly river, had to be designed to attain a considerable speed. The vessel was made in sections, put together and tested in London, where it attained a speed of over 13 miles an hour: it was then taken to pieces, shipped to India, refitted and tested with the following results, as stated in a letter from the Government Engineer.

Messrs. SHAND, MASON & CO.

Dear Sirs,—I have much pleasure in informing you that the Floating Fire Engine "Hooghly," is now fairly out of my hands. After repeated and very satisfactory trials, the vessel was made over to the Port Authorities for service, on the 28th ultimo. I had, altogether, six trials, both under weigh and pumping, each of which gave satisfactory results, and I am glad to be able to say *all* here are highly pleased with her performance, both in speed and throwing of water. The trials on the 20th ultimo and succeeding days, were with one, the 3 *inch*, six and twelve nozzles successively; average revolutions of pumps, 100; water pressure, 120 lb.; and steam pressure, 100 lb. on the square inch, throwing the water to a height of from 140 to 150 feet, to the admiration of all. It was rather an imposing sight for the residents of Calcutta. The vessel sped through the water with steam pressure at 100, averaging from 13 to 14 miles per hour, *nothing like it here.* I have had the "Hooghly" rather long on my hands, but this was owing to the great scarcity of labour. It is admitted by every one, that the "Hooghly" is a great acquisition to the port, and wishing that every success may attend her, I at the same time hope she will seldom have occasion to be required in action.

I am, dear Sirs, yours faithfully,
(Signed) W. H. SANDEMAN,
*Chief Engineer H.M. Dockyard, Calcutta.*

As a type of their DOCK FLOATING STEAM FIRE ENGINE, SHAND, MASON & CO. beg to refer to the London and St. Katherine's Dock Company, who have now in use, and in perfect order, one made for the Victoria (London Dock Company) in 1863, while the AUXILIARY FLOATING STEAM FIRE ENGINE is illustrated by that placed in a steam tug for the Surrey Commercial Dock Company (London), in 1867, and more recently by the small steam ferry boats on the Clyde, at Glasgow, being fitted with SHAND, MASON & CO.'S Patent Single Horizontal Steam Fire Engine.

The *North British Daily Mail* of 30th September, 1873, writes of the last as follows: "A Floating Steam Fire Engine to be used in extinguishing fires in vessels on the Clyde, was tested at the Custom House Quay yesterday afternoon. It is the same as the land Steam Fire Engine, minus the boiler, wheels, and carriage. The boiler that supplies steam for propelling the boat also works the engine. The jets used yesterday were 1 in., 1¼ in., and 1½ in.; the highest steam pressure was 48 lb. to the square inch, and the highest water pressure 120 lb., and it is estimated that the water rose to the height of 150 feet.

In addition to extinguishing fires in ships, this engine can be used as an auxiliary to the land ones in the event of a fire in warehouses at the Docks. Its superiority over the land engines in the case of shipping is, that it gets much closer to the fire, and draws its supply directly from the river, whereas on land, a proper supply can be had only by shutting the water off from the surrounding districts, and connecting it in the one in which the fire is raging, which sometimes occupies a good deal of time. The cost of this engine is not half the price of a Land Steam Fire Engine, and not the twentieth part of that of a complete Floating Steam Fire Engine similar to those in use on the Thames. The experiment yesterday was considered a great success, and was witnessed by a large number of people, among whom were the Hon. the Lord Provost, Bailies McDonald and Craig, James Deas, C.E., engineer to the Clyde Trust, &c."

SHAND, MASON & CO.'S SELF-PROPELLING FLOATING STEAM FIRE ENGINE, as constructed for the London Fire Brigade.

SHAND, MASON & CO.'S SELF-PROPELLING FLOATING STEAM FIRE ENGINE, as constructed for the British Government.

For Floating Steam Fire Engines to which reference may be made, see page 14.

SHAND, MASON & CO., FIRE ENGINE MAKERS AND HYDRAULIC ENGINEERS,

*Perhaps the world's first steam-propelled, steam-pumped fireboat was this Shand Mason & Co. self-propelling floating fire engine built for the London Fire Brigade in 1855. The craft fought every fire along or near the River Thames for a quarter of a century. From the Collection of Bill Dahlquist*

own individual preferences and estimate of the necessities would have put the $105,000 or thereabouts which *The New Yorker* cost, in two boats. Unquestionably, that would have furnished greater firefighting facilities and fire protection for the city.

"Pioneers in any specialty will understand what I mean when I say that a new thing has to be introduced with some tact; you cannot always go the whole figure. You have often got to cut your cloth according to the dictation from the powers that be." Politically pragmatic, Cowles realized the customer may be wrong-headed, but is nevertheless the one paying the bill and, therefore, makes the ultimate decision."

There was no arguing the point, however, that *The New Yorker* packed a mighty wallop and was the forerunner of even more powerful fireboats. Society members noted the FDNY's other boats: the 2870-gpm *Havemeyer,* named after a mayor, and the *Zophar Mills* rated at 2400 gpm and honoring the legendary foreman of Eagle Engine Company No. 13 of the historically illustrious New York Volunteer Fire Department. With *The New Yorker* as the flagship of the fleet the three boats mounted a formidable attack.

"When the gong rings the boats approach the fire as close as they can," said Parsons. "The men are all protected. Those who maneuver (*The New Yorker*) are in the pilothouse and behind the metal shields. No men are left on the decks except those who have to handle the hose. There is one hose always in readiness in case the boat itself catches fire. Last spring I put in a sprinkler system on *The New Yorker* so that the boat and men are continually drenched during the time they are fighting a fire.

"Having fastened alongside of a dock or wharf," said Parsons, "They turn on a heavy stream and knock a hole in the building, so that the men and hose can go right to where the fire is. If the fire is very hot, they will throw a heavy stream from their monitors and the men can get behind the shields. If they want to change the direction of the stream, they can do so very quickly (through swiveled nozzles)."

Cowles recalled that "when fireboats were first mooted— I mean modern fireboats—there were no shore engines that could throw a 2-inch stream of water 200 feet and keep it up…Streams above two inches in diameter *were not known.* Streams that would tear their way through a warehouse, strip the joiner work off the deck of a ship, tear a lumber yard pile to pieces and throw it off the dock—had not been dreamed of. These fireboats did it."

Cowles qualified his comments by noting the precedent for these large caliber, high-pressure streams originated during the Gold Rush era when prospectors used nozzles, often stationary and mounted on pylons, to jet powerful streams which washed away entire hillsides to more quickly expose veins of ore.

*Superintendent James Braidwood of the London Fire Engine Establishment took a profound interest in developing fireboats to protect the high-value River Thames waterfront. Ironically, he was killed at this waterfront fire when he was buried under a collapsed wall while a London fireboat supplied hoselines to firefighters battling The Great Tooley Street Fire of 1861. From the Collection of Paul Ditzel*

"I do not believe there are many buildings facing the river that *The New Yorker* could not demolish, and there are very few ships of war whose men *The New Yorker* could not drive from their guns if she were permitted to get close enough," said Parsons. "The four-inch stream, which is the largest *The New Yorker* throws today, has demolished a pile of paving stones and has washed them along the street as if they were marbles...although at a greater distance than is required in fighting a fire."

Parsons' simile between fireboats and warships caught Admiral Richard W. Meade's attention. He suggested fireboats could serve as auxiliary defense weapons. In an unstated but obvious reference to labor strife that was endemic in the nation, Meade said:

"I remember that in the late great unpleasantness I commanded one of the ironclad on the Mississippi River in 1862...and we had a couple of hot water arrangements to keep the enemy from boarding. We had four men on each side of the vessel to attend the hot water hose. We could throw a stream of scalding water in a few seconds. Although I have known men bold enough to board in the face of grape and canister (shot) I have never seen any set of men bold enough to board in the face of scalding water."

The suggestion that fireboats would be useful for shooting scalding water at workers demanding better job conditions, including the eight-hour-day, was repugnant to society members who quickly brought the subject back to waterfront fire protection. In addition to fighting fires the fireboats proved their worth as floating pumping stations to supply shore fire engines and water main systems in high-value districts well inland from dockside.

While fireboats were effectively demonstrating their capabilities for waterfront blazes and as floating pumphouses supplying apparatus nearly a mile away, that acceptance had been a long time in arriving, especially in American cities. Cowles recalled that only a decade earlier, "It took a man who was willing to stand and able to bear the 'slings and arrows of outrageous fortune' to go before city fathers and officials and preach fireboats."

"It was an uphill battle," said Cowles, referring to the years preceding 1896. "The fireboats that have been put in service in the last seven, eight or nine years have shown beyond question (that they are) the only proper way to furnish fire protection for waterfront belts of marine cities. That area you will find in almost every marine city in this country covers practically the (entire) storage manufacturing and business portion."

The slings and arrows of disparagement to which Cowles referred must be put in context to appreciate the spurt of fireboat interest starting around 1875 and peaking by 1896 and the years immediately thereafter. There were fireboats starting around the mid-1700's, notably in London and other highly-developed European ports. London's first two fireboats, no doubt typical of those elsewhere, were primarily intended for fighting fires in large warehouses on the banks of the River Thames.

Up to 1852, a turning point as we shall shortly see, the boats were barges with a large hand-operated fire engine on each. One of the engines was stroked by 40 men and the other by 60. These fireboat firefighters, officially designated as pumpers, were paid one shilling for their first hour's effort and 2½ shillings an hour after that. Obviously, this was not the most desirable of exertions, so there was a bonus: Pumpers were entitled to as much free beer as they could drink. If, however, the beer ran out as might happen during a prolonged fire, the pumpers walked off the job.

In 1852, Superintendent James Braidwood of the London Fire Engine Establishment stood both furious and frustrated as the beer supply and the pumpers simultaneously ran out during the burning of Humphrey's Warehouse on the River Thames. (There is a tragic postscript to the Braidwood story. He was killed in 1861 when a wall collapsed during the Great Tooley Street fire on London's waterfront.)

The Humphrey's Warehouse blaze finally tipped the scales in favor of steam fire engines, which did not drink beer, but continued to chuff so long as their boilers were being fed coal. Soon after that fire London first introduced steam into the brigade with the purchase of a steam-operated fire engine on a barge. Three years later, in 1855, a powerful steam-propelled and pumping engine was constructed and the manual floating engines became a memory.

Compared with fireboat technology in England, the United States lagged by more than half a century. While American ports were far less industrialized than those in Europe, there nevertheless existed a need for protection, even in colonial times and later, when waterfront areas frequently burned because they were not accessible from the landside. There was, moreover, the matter of cost and constant maintenance, both problems in an American fire service which was wholly volunteer until immediately after the Civil War.

Probably adapting London's first and least costly approach, New York volunteer firemen built America's first fireboat in 1800. They mounted a hand-operated pump aboard a scow with a sharply pointed bow and a square stem. *The Floating Engine* (also known as *The Floater* or *Engine Company No. 42*) was rowed to fires by the volunteers from their berth on the East River at the foot of Roosevelt Street.

Amazingly, they had enough strength remaining to operate the pump when they arrived at the fire. The pump was of the coffee mill type. Its front and rear pump handles were rotated, much like an old time coffee grinder. *The Floating Engine* saw 18 years of service and probably was most often used to supply water to land-based fire engines which were also operated by hand. *The Floating Engine* had no fixed nozzles and its pumping capacity probably was no more than several hundred gallons a minute.

Little known is the significance of the Civil War in furthering fireboat technology. John Braithwaite, an inventor, had vainly tried to interest London and New York insurance companies in a steam-powered pumper named *Novelty* that could put out 250 gpm. Ignored, little was heard of him until the Civil War. With John Ericsson he constructed the ironclad cheesebox-on-a-raft *Monitor* that fought the standoff battle in Hampton Roads, Va., March 9, 1862, with the Confederate *Virginia*, which was mostly crewed by Norfolk volunteer firemen.

The Civil War spawned a host of designers and builders of boats that one day would replace clipper and other ships. Seeking new markets for their designs, some builders' ideas were bizarre mostly because they saw little difference between a warship battling the enemy and a fireboat bat-

*America's first fireboat, **The Floating Engine**, was built in 1800 by New York City volunteer firemen. Mounting a coffee mill type of hand-operated pump amidship,* *the volunteers rowed to fires. The fireboat probably supplied no more than several hundred gallons of water a minute to land-based hand-pumpers. FDNY*

tling a fire.

Perhaps the most curious and most-publicized of these proposed fireboat contraptions was that of Sir Hiram Stevens Maxim, a native of Sangerville, Maine, who is best-known as the inventor of a machine gun bearing his name, smokeless powder, and a heavier-than-air plane. Maxim may have been a whizbang with machine guns but he never bothered to learn basic fireboat requirements.

Had it been built, Maxim's fireboat would have been the world's longest: 250-feet; twice that of *The New Yorker*, which even its designer believed to be too long for operating effectively around waterfronts. The size of Maxim's boat would, therefore, have provided as much maneuverability as an ocean liner in a duck pond.

The boat was designed with port and starboard paddlewheels amidship and aft-mounted boilers with twin stacks. Except for the almost total lack of a super-structure it resembled a Mississippi River steamboat. Near the bow, Maxim proposed to mount a tall water tower topped by a crow's nest and a nozzle itself dozens of feet long. Reflecting contemporary naval gunnery, Maxim's monitors were configured like cannons and, similarly, swiveled on trunnions. The monitors' discharge openings were 20-inches in diameter: five times larger than the mighty *New Yorker's*.

After much newspaper and magazine fanfare, Maxim's fireboat idea faded into oblivion. Perhaps because of sug-

gestions like these, Cowles and his fellow naval architects were easy targets for those slings and arrows he mentioned.

Fire chiefs' reluctance to accept proposals for boats especially built for firefighting is partly explained by their traditionally conservative attitudes. The history of American fire apparatus evolution is replete with resistance to innovations. Steam-operated fire engines were long shunned by volunteer firefighters who cherished their hand-operated pumps. That changed with the gradual establishment of full or partly-paid fire departments starting after the Civil War. Following the steam engines' acceptance, the American fire service was slow to convert to motorized apparatus.

No doubt the major objections to modern fireboats, however, were their limited purpose and high costs of buying and maintaining them. Tugboats and fire engines on barges surely had their drawbacks; not the least of which were their undependable immediate availability and low pumping capacities. There seemed, therefore, to be justification for accepting these calculated risks when costs of modern fireboats were scrutinized.

Modern fireboats had voracious appetites for coal which was consumed around-the-clock to keep boilers near pumping pressures. Frequently mentioned was the fact that fireboats burned more coal at their berths than they ever did at a fire.

*Boasting water cannon with 20-inch diameter openings—five times larger than any other fireboat—this 250-foot-long craft was proposed in 1881 by Sir Hiram Stevens Maxim, best-know as a machine gun inventor. The boat received lots of publicity, but no interest whatsoever among fire departments. From the Collection of Paul Ditzel*

Make no mistake about it, the cost of fireboat upkeep, and labor costs of fireboat firefighters idling away hours and days between alarms offered compelling arguments against modern fireboats. Add to that the costs of building and operating waterfront fire stations to house firefighters. Few boats had on-board facilities for eating and sleeping; New York's *William F. Havemeyer* being a notable exception. This boat was crewed by a captain, two lieutenants, a pilot, two marine engineers, and five firemen.

But a new economic era was dawning in the United States and the fire service was ill-prepared for the concomitant challenges. These developments started after the Civil War and escalated into full-blown economic growth which two severe financial depressions could not stifle. The 1869 completion of the first transcontinental railroad marked the opening of the west for economic expansion.

Concurrently, America quickly transitioned from predominately small business enterprises to heavily-industrialized corporate empires from the northeastern seaboard west to the Great Lakes. Steel, grain, flour milling, meat packing, lumber, warehousing, tanneries, huge wooden bunkers for coal storage, flammable liquids storage, and a diversity of manufacturing complexes jam-packed waterfronts. Populations of cities doubled and tripled especially where networks of railroads met waterways. Baltimore, Boston, New York, and Philadelphia became not only industrial hubs but export-import centers, too, as American products exerted a strong impact upon foreign trade.

With this historically unparalled growth came myriad fire problems. In the haste to put up commercial structures, many of them wooden buildings on wooden piers and wharves, fire codes were ignored, if indeed they existed. The inevitable result was a rash of major fires, many of them on or near congested waterfront and adjacent districts where access by shore-based land apparatus was difficult or impossible. "The Scientific American" reported, March 5, 1881, that "fully half of the fires in New York and Brooklyn are confined to the riverfront." Similar problems existed in other cities with busy ports.

Fire chiefs and other city officials realized the conflagration breeders along their waterfronts and nearby commercial areas. On October 8, 1871, fire started in a tinderbox residential district west of the Chicago River spread into the city's industrialized complexes along the river. Brick warehouses were consumed almost as fast as lumberyards and furniture factories. Firebrands hurtled the narrow river and touched off more buildings in the downtown district.

Fire Chief Robert A. Williams had earlier petitioned city

*Tugboats leased by many cities on an as-needed basis were the chief line of defense for waterfront cities. Los Angeles, for example, leased these steam and sail-equipped tugs, the **Warrior** and the **Falcon,** shortly after the city annexed San Pedro and Wilmington in 1909, and the fire department assumed waterfront protection responsibilities. From the Collection of Harry Morck*

*Fire engines on barges were interim solutions by some cities unable to afford modern fireboat protection. Starting in 1917 Los Angeles used these Amoskeag and Nott steam fire engines. Horse-drawn from their station and loaded on the barge, they were towed to fires. Note bags of coal stored aboard barge. From the Collection of Paul Ditzel*

officials for a fireboat but was told privately-operated tugs were more cost efficient and could meet any firefighting need. Not that a modern fireboat could have stopped The Great Chicago Fire, but at least it would have provided protective water curtains against those hurtling firebrands while supplying land companies which could not find sufficient pressure in the mains to produce effective fire streams. Chicago's downtown area, including City Hall, was devastated. The fire then leaped the Chicago River again and continued its rampage. Hundreds of factories and other commercial buildings were destroyed as were 17,450 homes. Around 300 people were killed, and damage was around $200 million.

Almost exactly one year to the day later, November 9, 1872, a fire starting in Boston's mercantile district swept into the waterfront and destroyed at least half a dozen piers and all nearby structures. The 18-hour fire took 776 buildings, killed 15 firefighters and caused $75 million in damages. Ironically, Boston Fire Chief John S. Damrell had received approval to buy a fireboat, but it was several months from completion when the conflagration occurred.

Boston firefighters therefore did what they could to check the stampeding onslaught including use of hoselines supplied by a tugboat, the *Lewis Osborn.* Two months later, America's first steam fireboat, the *William M. Flanders,* went in service, January 1, 1873, at Boston's Central Wharf. The 75-foot *Flanders* was outfitted—with four Amoskeag pumps delivering a total of 2500 gpm.

The *Flanders,* displacing 55 tons, was designed by Damrell and had an iron hull. Fireboat longevity, usually long, had a relatively short service life as far as the *Flanders* was concerned, and the craft was put on reserve duty after only 16 years. Replacing the *Flanders* was the *John M. Brooks,* an 108-foot-long boat with a beam of 24-feet and a 9-foot-draft. Its rated pumping capacity—4240 gpm at the very minimum—was more than double that of the *Flanders.*

Footnoting fireboat history is the fact that many of these craft were acquired shortly after a waterfront conflagration dramatized the need. On June 6, 1889, the entire waterfront of Seattle, a fast-growing lumber port was destroyed along with 31 blocks of high-value commercial district. The following year Seattle got its first fireboat, the *Snoqualmie,* named after a well-known Indian tribe in the Pacific Northwest. The 5765 gpm *Snoqualmie* was wood-hulled, 90-feet-long, with a 22$\frac{1}{2}$-foot-beam and a 10-foot draft.

The alarming increase in major fires along waterfronts and adjacent business and commercial districts magnified the need for greater volumes of water. That this problem was outstripping the capability of cities to provide it was shown in high-value districts, which were steadily more crowded with structures containing high fireloading, and as manufacturing and other commercial activities became more hazardous. Buildings, moreover, were growing taller as downtown land values demanded high density usage.

Further increasing the need for more water were apparatus manufacturers' newer steam fire engines rated in excess of 1000 gpm. In 1874, moreover, a New Haven, Conn., piano factory owner, Henry Parmelee, devised the first practical automatic fire sprinkler head and installed a protective sprinkler system in his plant.

*Among the first two dozen modern fireboats built was the **John M. Brooks** with a pumping capacity rated at 4240 gpm. This Boston fireboat mounted two deck guns forward and went in service in 1889 to replace America's first steam fireboat, the **William M. Flanders.** From the Collection of Bill Noonan.*

*A year after Seattle's waterfront area was destroyed by a conflagration, June 6, 1889, the city purchased its first fireboat, the **Snoqualmie**, named after an Indian tribe in the Pacific Northwest. The boat had a 5765 gpm capability, was 90-feet-long and wood-hulled. From the Collection of Steven Lang*

Sprinkler systems required still more water, as did the advent of firefighting water tower apparatus, starting November 8, 1879, during a three-alarm fire at 82-84 Bush Street in New York's Greenwich Village. These water towers, subsequently acquired by many fire departments, had no pumping capability and depended upon engine companies or fireboats to supply large volumes of water to the elevated nozzles.

Not the least of the concerns over inadequate water supplies were those expressed by fire insurance underwriters alarmed by a steady increase in the number and severity of large loss high-value district and waterfront fires.

Starting with Rochester, N.Y., in 1874, some cities installed underground high pressure water supply systems as a firefighting supplement to the traditional water main gridwork and hydrant systems around waterfronts and adjacent high value districts. In Buffalo, for example, an alarm from an area encompassed by the high pressure system required pumping station workers to engage pumps which, drawing upon the city's Lake Erie supply, charged the system until the chief commanding the fire ordered the pumps shut down.

Detroit's system was further illustrative. Thirteen pipelines, each nearly one mile long, extended from waterfront inlets to deep inside the city's high-value district. These 8-inch-diameter pipelines fed special hydrants located no more than 300 feet apart. Each hydrant was not only equipped with 3-inch controllable outlets, but a telephone jack for the fire chief's use in talking directly to the captain of the 115-foot fireboat, the *Detroiter*, a wood-hulled boat built in 1893 with a pumping capacity exceeding 5000 gpm.

Parsons told society members how the *Detroiter* demonstrated its mettle by pumping into the pipeline feeding a hydrant and hose which arced a two-inch stream over the top of a 10-story building 2000 feet from the riverfront. During a fire it pumped non-stop for 19 hours while supplying half-a-dozen 3-inch streams through two hydrants and three lines of 3-inch hose directly from the boat to shore-based engines.

This boat hose was carried on reels aboard fireboats and often in boat tenders: horse-drawn hose wagons which answered alarms in high pressure water system districts. Later motorized, these tenders often mounted a fireboat-sized cannon atop the rigs.

Despite the need for boats especially built for firefighting, many cities continued to make-do with traditionally cost-effective approaches. In 1866, immediately following the Civil War when the Fire Department of New York replaced the old volunteer system, the city leased Baxter Wrecking Company's steam tugboat, *John Fuller*, and installed two Amoskeag pumps delivering 2000 gpm. Not until nine years later did the FDNY acquire its first modern type fireboat, the *William F. Havemeyer*, a 106-foot-long boat rated at 2870 gpm. Historical records indicate New York may have bought the *Fuller*, too.

As the meeting of the fireboat designers ended, they optimistically anticipated more assignments that were certain to come during this first heyday of the fireboats. Strangely, they seemed only slightly impressed by the significance of their discussion, or Parsons' presentation, which would one day serve as a benchmark for fireboat designers, builders and historians.

Parsons' treatise was dismissed in the society's minutes as "a very innocent paper", as members went on to the next presentation, "America Corn-Pith Cellulose." Not one of those fireboat pioneers could have had the slightest notion of the imminent horrendous events which would put extraordinary demands upon their new breed of fireboats and forevermore cement their place in the American fire service.

Boston's **Angus J. McDonald** was among the first 25 modern fireboats built and saw more than half-a-century of service. Built in 1895 the wood-hulled fireboat originally had a pumping capacity rated at 5765 gpm. The **McDonald** was 110-feet-long, with a beam of 26-feet and a draft of only 8-feet 6-inches. The boat displaced 178 tons. More than any other fire apparatus, fireboats usually have lengthy service lives. Well before her 1947 retirement the McDonald battled this four-alarm fire, November 7, 1933 at Boston's Central Wharf. From the Collection of Bill Noonan

Milwaukee's first fireboat, **Cataract**, was built in 1889 and had a capacity of 4660 gpm. Displacing 245 tons, the wood-hulled boat was 106-feet, 9-inches long, had a 24-foot beam and a draft of 9-feet, 6-inches. From the Collection of Bill Dahlquist

The **Yosemite**, built in 1889, was capable of over 5000 gpm. The 109-foot-long boat with a beam of 24-feet and a 11-9-inch draft, displaced 144 tons. In 1894 the **Yosemite** foundered off Chicago's 75th Street during a Lake Michigan gale. Raised and rebuilt, the **Yosemite's** name was twice changed: **Protector** (1903) and **Michael J. Conway** (1907). From the Collection of Robert Freeman

White was a favorite color for many of the first modern fireboats, including Phila-delphia's **Edwin S. Stuart.** Built in 1893, the Stuart delivered 5765 gpm; making it one of the most powerful of the new breed of fireboats. The Stuart was steel-hulled, 105-feet-long, with a beam of 23-feet, 7-inches and a 9-foot, 9-inch draft. Its top speed was 10 knots; typical of the first boats especially built for firefighting. From the Collection of Bill Dahlquist

Cleveland's steel-hulled, 90-foot-long, **John H. Farley,** was the city's first fireboat. It had a beam of 22-feet and a draft of 7-feet, 6-inches. Built in 1894, the same year as its second boat, the **Farley** had a pumping capability of 3360 gpm. From the Collection of Steven Lang

Note the unusually large turret gun forward on the wheelhouse of Milwaukee's wood-hulled **James Foley,** the city's second fireboat. Delivered in 1893, the boat was among the most powerful of the first modern craft built and could pump 5010 gpm. The **Foley** was 107-feet -long, with a beam of 24-feet and an 8-foot, 6-inch draft for operating in the city's many shallow waters. From the Collection of Bill Dahlquist

Postcard representation of Cleveland's second fireboat, **Clevelander,** battling a waterfront blaze. The wood-hulled craft, built in 1894, was 82-feet-long, had a beam of 21-feet, a draft of 8-feet, 6-inches and offered a pumping capacity of 5010 gpm. From the Collection of Steven Lang

# CHAPTER THREE
## THE FIREBOATS THAT SAVED CHICAGO

At 4:09 that sultry Saturday afternoon, June 30, 1900, the fire alarm register began to ping in the headquarters office of Edward F. Croker, chief of the Fire Department of New York. The register's tapper struck the bell twice, paused, hit five more times, and hesitated before tapping the bell once.

Box 251. West and Morton Streets on the North (Hudson) River. Realizing this was a high-hazard district in lower Manhattan and along the waterfront as well, Croker left immediately, as he always did when alarms were sounded from these areas. Also answering the alarm were two fireboats, *The New Yorker*, from its Castle Garden (Battery Park) berth at the southerly tip of Manhattan, and the *Zophar Mills*, berthed at Manhattan Pier 42 on the North River.

Directly across the river from Box 251, firemen saw mammoth clouds of black smoke and flames engulfing four long piers and ocean liners of the North German Lloyd Company terminal in Hoboken.

Apparently starting among cotton bales on the 800-foot-long main pier, the fire had ignited a nearby cargo of whiskey. With incredible speed, the fire swept aboard four of the world's best-known ocean liners, which were teeming with activity while taking on passengers, stores and cargo prepatory to departure for Germany.

The 500-foot-long *Bremen*, only four years old, and registered at 10,000 tons; the famous *Kaiser Whilhelm der Gross*, the fastest and one of the most popular liners afloat; the *Main*, a new steamship of 6398 tons which was about to complete her maiden voyage, and the liner, *Saale*.

Explosions cast clods of blazing cotton far out onto the river and set a fire on Ellis Island, while piers and sheds and other combustibles around the North German Lloyd terminal, together with coal and supply barges, burst into flames. Even as Croker reached the waterfront, he saw tugs pulling the blazing *Kaiser Wilhelm* and *Bremen* clear of the timbered piers.

Ships' crew, stevedores and passengers were leaping overboard. Many of them were shrouded in flames. The fully-involved *Bremen* and *Saale* also were being nudged from their berths by tugs while Croker watched the backdrop of thick clouds of darkening smoke testifying to the worst catastrophe up to that time in or near the Port of New York.

Croker signaled *The New Yorker* to pick him up. He often used this flagship of the New York fireboat fleet to direct waterfront firefighting. As the *Zophar Mills* approached, Croker called for the *William L. Strong* from its Grand Street and the East River berth to join what would be a rescue and firefighting operation without parallel up to that time in the history of American fireboats. Even with the combined firepower of the boats' more than 22,400 gpm, there was no avoiding catastrophe as the fireboats plowed into the fiercely radiant heat.

*The New Yorker's* berth at Castle Park (Battery Park) at the southerly tip of Manhattan, enabled the fireboat to quickly answer alarms such as those for the Hoboken, N.J. and **General Slocum's** disasters. **The New Yorker's** crew slept aboard the boat until the station was completed in 1895. The hose-drying tower was popularly referred to as The Chinese Pagoda. Castle Park also was an amusement park with an aquarium ( left of the fire station) as well as an immigration center along with Ellis Island. The historically-famed area was razed in 1941 with the start of construction of the Manhattan-Brooklyn Tunnel. From the Collection of Steven Lang

The North German Lloyd ocean liner and pier disaster June 30, 1900, in Hoboken, N. J. caused great mourning in Germany because many of its citizens were among the more than 300 people killed. Postcard illustrations of the burning ocean liners, **Main** and **Bremen**, more graphically showed the holocaust than contemporary photos. From left to right the New York fireboats portrayed probably are **The New Yorker** the **William L. Strong** and the **Zophar Mills** although the artist's conceptions markedly differ from the actual profiles of the fireboats. From the Collection of Steven Lang

The **William L. Strong's** 51-year career started shortly after its 1898 commissioning when it was called to battle the North German Lloyd ocean liner and pier disaster in 1900 and the **General Slocum** catastrophe four years later. Designed by Harry deBerkeley Parsons, the steam-powered **Strong** weighed 203 tons, had a 100-foot keel and a 6-foot draft. Its capacity was rated at 6500 gpm at 150 psi. The boat mounted three monitor guns. Construction cost was $56,490. From the Collection of Robert Freeman and Ray Henry

Many passengers, crew and stevedores died below deck of suffocation and burns as the rapidly-spreading smoke and flames blocked their escape. Wherever Croker looked there were people screaming for rescue amid silent, floating bodies.

*The New Yorker's* crew, using their protective metal shields against heat, put the boat alongside the blazing *Bremen* and pulled 28 people to safety through the portholes. Dozens more died because other portholes were too small or red hot to squeeze through.

Later, when there were no more rescues to be made, the fireboats focused their guns on the blazing hulks while firefighters continued to search the East River for bodies. Many were never recovered. Reliable estimates put the toll at between 326 and more than 400 dead. Property damage to the ships and piers exceeded $5 million.

A calamity many times worse would shortly eclipse the North German Lloyd fire. Before that, however, a conflagration devastated the commercial and waterfront district of Baltimore. On a quiet Sunday morning, February 7, 1904, a fire was discovered in a downtown building at 10:48. Flames swept through tall brick buildings in the mercantile area and, by nightfall, were invading the waterfront.

Baltimore's only fireboat, the *Cataract*, supplied water to steam fire engines and, later, darted in and out of the smoke and heat while using her four guns to pinch off flames involving waterfront structures and lumber yards. The nine-year-old *Cataract*, rated at 4400 gpm, was little match for the flames which rampaged until the next morning. The Great Baltimore Fire decimated 140 acres—70 blocks of buildings—and caused losses estimated at between $45 and $100 million. Luckily, no lives were lost.

*Baltimore's **Cataract** was 78 years old when it was replaced in 1969. The city's first fireboat's unique claim to fame was its battle against the Great Baltimore Fire in 1904. Although remodeled over the years, the **Cataract's** profile remained basically unchanged, except for the addition of an elevated steel tower mounting a turret gun. Note the many hose supply inlets just under the base of the bow turret platform. When built in 1891, the **Cataract** had a 4400 gpm pumping capacity rating; was 85-feet-long, an 18½-foot-beam and a draft of 9-feet. Charles Cornell*

No such miracle awaited more than 1000 people—mostly children and women—who crowded aboard the side paddlewheel excursion steamer, *General Slocum*, shortly before 9:30 Wednesday morning, June 15, 1904, at the East 3rd Street Pier on New York's East River. For the 740 children aboard, today was an outing for members of St. Mark's Lutheran Church. The short trip up the river and past Hell Gate was to take them to a picnic grounds in the Bronx. The 264-foot-long triple-decker left the dock at 9:40 a.m. Minutes later, a small fire, feeding upon many layers of paint and varnish, quickly spread throughout the wooden ship.

Instead of immediately beaching the *Slocum* to give everybody a chance to escape and firemen a fighting chance to save the ship, Captain William H. Van Schaick ordered full speed ahead while looking for a deserted area to run her aground. The *Slocum's* speed, together with the river breeze, worsened the flames. As the vessel steamed past Hell Gate, telephone reports of the fire began to be received in the Fire Department of New York alarm office. The locations given were vague, but almost immediately Box 2339 was pulled at 10:13 a.m. The firebox was on the Manhattan waterfront across from Hell Gate. Croker answered the alarm.

The *Zophar Mills* was the nearest fireboat. Its crew was dumbstruck as they left their berth and saw the *Slocum*, trailing smoke and flames. Women and children, their clothing afire, were leaping overboard. With the *Zophar Mills* giving chase, Croker ordered the *Abram S. Hewitt* from its Brooklyn berth to scoot across the river and pick him up at the foot of 67th Street, while the *William L. Strong* joined the fireboats and tugs steaming toward the blazing *General Slocum*.

Van Schaick finally beached the vessel off Locust Island (near Rikers Island) and the fireboats moved in on the steamer. By then the *Slocum* tinderbox was little more than a hulk concealing hundreds of bodies and dozens more wedged in its paddlewheels. The final death toll was put at a minimum of 1030. It was the worst disaster in New York City history. Convicted of manslaughter, Captain Van Schaick, 67, was sentenced to 10 years in prison and probably would have died in Sing Sing, except for his pardon in 1911 by President William Howard Taft.

The Hoboken and *Slocum* disasters reinforced Croker's interest in fireboats. Since becoming chief in 1899, Croker often spent Sundays visiting fireboats and developing tactics which were to lead to highly-specialized techniques in the craft of waterfront firefighting. Recognizing this new specialty, Croker created, June 16, 1905, a marine battalion consisting of officers and crews of the FDNY's fireboat fleet. Croker could not have chosen a more qualified fleet commander: Battalion Chief John Kenlon, the department's only officer who held a master's and a pilot's license.

The idea of a firefighting navy to augment land-based companies proved so viable that Croker, in 1909, created the FDNY Marine Division and promoted Kenlon to deputy chief as its commander. The Marine Division was a first in American fire service history, and its feats were heroically spectacular during a period spanning more than 80 years.

The **Zophar Mills**, named after one of New York's legendary volunteer firemen, was a veteran of the Hoboken and **General Slocum** disasters. The **Mills** is shown in this historically intriguing photo as a side paddlewheel excursion steamer approaches (extreme left) that has the identical profile of the **Slocum**. It is not known whether the paddlewheeler was the **Slocum**. The steel-hulled **Zophar Mills** was built in 1882 and was 125-feet-long, a few inches shorter than **The New Yorker**. Its Amoskeag and Clapp & Jones fire pumps were rated at a minimum of 2400 gpm. From the Collection of Bill Dahlquist

Among the New York fireboats racing toward the blazing **General Slocum** was the nearly new **Abram S. Hewitt**, which Fire Chief Edward F. Crocker boarded to take him to the scene of the disaster. The **Hewitt's** designer, Harry deBerkeley Parsons, supervised construction of the $83,750 fireboat rated at 7000 gpm at 150 psi. Built in 1903, the **Hewitt** had three monitors and twin smokestacks; one of which was later removed. Steel hulled, the fireboat had a 117-foot keel, a 25-foot beam and a 10½-foot draft. Its single screw propeller drove the boat a maximum of 10 knots. The **Hewitt** was built at the New York Shipbuilding Co., Camden N. J. From the Collection of Steven Lang

Postcard producers often capitalized upon major news events, especially when actual photos failed to capture the fire spectacle. This 1904 postcard appeared shortly after the **General Slocum** disaster on June 15 of that year. Except for the fairly good representation of **The New Yorker**, the blazing excursion steamer lacks the **Slocum's** side paddlewheels. Apparently to hype the scene, the artist chose to show the burning steamer under a full moon. The disaster occurred during a sunny morning. From the Collection of Steven Lang

Croker was as keenly interested in developing water-front firefighting techniques as he was in the technology of fireboats. When New York approved his recommendation for three more fireboats, he outlined his ideas:

"The fireboat that I would recommend should be about 120-feet-in-length overall, with a beam of 24 or 25 feet and a draft of not more than nine feet and should be built of steel. The decks should be flush fore and aft and as clear of obstructions as possible. The common practice of providing a large pilothouse should be avoided," Croker wrote. Smaller pilothouses he observed offered better visibility.

Croker's concepts were considerably at variance with those of the naval architects who expounded their ideas only a decade earlier during that noteworthy meeting in 1896 of the Society of Naval Architects and Marine Engineers. They were largely responsible for the design of the first 25 boats especially built for firefighting. A comparison of Croker's ideas with those of the naval architects does not suggest that either viewpoint was right or wrong, but illustrates the rapidity with which fireboat technology was advancing. The conflicting ideas, moreover, presented a blend of naval architectural theory with pragmatic water-front firefighting experience.

The designers had deprecated fireboats as long as those Croker envisioned. They felt lengthy boats were far less maneuverable in narrow slips and posited that two smaller boats would be far more effective than a 120-foot fireboat espoused by Croker. Fifteen of their 24 boats were wooden hulled. Croker preferred steel, both for sturdiness and longer life.

The naval architects recommended 6000 gpm as the maximum capacity. The fire chief chose to increase that to a minimum of 8000. Croker's preference for twin propellers over a single propeller was predicated upon better maneuverability and speed. The architects concurred that twin propellers were preferable but claimed their advantages were outweighed by space and mechanical considerations.

Croker's flush deck concept was logical because firefighters could more easily and safely move about. He also urged more fixed monitor guns than most earlier fireboats; some of which mounted only one. The chief also chose to increase the diameter of the guns and to provide more water manifold outlets so additional hoselines could supply land-based fire engines than was customary with most conventional boats.

Perhaps recalling the ineffectiveness of monitor streams being lobbed into the burning ocean liners during the Hoboken disaster, Croker recommended an aft-mounted tower mast topped with a turret gun. From that height a powerful stream could be jetted into upper decks of ships or topmost floors of buildings.

Even more revolutionary were Croker's recommendations for propulsion and pumping:

"If the turbine could be reversed as quickly and as certainly as the reciprocating engine (universally used in all

New York's **Thomas Willett** mostly epitomized Fire Chief Edward F. Croker's concepts for a new breed of fireboats. As with its sister boat, the **James Duane**, the **Willett** was the first of the new fireboat genre to be equipped with steam turbine driven centrifugal pumps rather than the traditional steam piston reciprocating engines. Croker's desire for better visibility from the wheelhouse is forcefully illustrated as is his pioneering concept of an elevated steel tower with a monitor gun. The **Willett's** 4500-9000 gpm capacity is shown as it pumps into a land-based water tower apparatus. Note that the **Willett's** high pressure, probably in this case about 300 psi, is bowing the water tower's elevated mast. From the Collection of Robert Freeman

other fireboats of that era) I would favor it, but in this regard many things remain to be demonstrated. I recommend that the reciprocating engine be used for propulsion, and the turbine to operate the pumps," said Croker.

"Centrifugal pumps, when turbine driven, will do the same work with half the steam required by reciprocating units. I therefore believe that centrifugal pumps having a capacity of at least 4000 gpm be installed and that they be driven by turbine engines."

Two of the three boats New York authorized incorporated Croker's ideas: the *James Duane* and the *Thomas Willett*. New York obviously hedged its bet on Croker's concepts. The third boat, the *Cornelius W. Lawrence*, was mostly conventionally built with reciprocating pumps and without an elevated tower.

History would demonstrate, however, that fireboat technology was entering the twilight of reciprocating engines, and the dawn of turbine driven centrifugal pumps, elevated water towers, and more and larger diameter guns. The result: greater volumes of water and higher pumping pressures that became necessary as waterfront firefighting problems escalated concurrent with the size of buildings and related hazards.

The *James Duane* and *Thomas Willett* were built at the Jersey City, N. J., yards of Alexander Miller & Brothers at a cost of $118,925 each. As fire chiefs have always known, budget and other constraints mandate compromises in apparatus purchases, and these fireboats were no exception. The steel-hulled boats had single propellers, contrary to Croker's preference, but otherwise closely paralleled his recommendations. Each was rated at 9000 gpm at 150 psi, when large volumes of water were needed, or 4500 gpm at 300 psi, when higher pressure was called for to penetrate the heart of large fires or areas the boats could not easily reach.

This new generation of fireboats each mounted two bow monitors with 3-inch diameter nozzles. A third gun was aft-mounted on a stationary tower of steel gridwork. Although both boats were commissioned December 7, 1908, the *Duane* was delivered first and therefore is often cited as the first fireboat with an elevated tower.

That claim must be qualified. Four years earlier, Portland, Ore., took delivery in 1904 of its first fireboat, the *George H. Williams*, which had an elevated standpipe coupled to a monitor nozzle, but with no steelwork supports as aboard the New York boats. It can only be speculated that Fred A. Ballin designed the 6200-gpm *Williams* with the standpipe which extended well above the twin smokestacks for fighting frequent fires aboard tall lumber schooners plying the Pacific Coast, or the sprawling lumber yards and mills along the waterfront. Those fires could more effectively be hit by a straight stream from the standpipe nozzle than by those lobbed by the *Williams'* four deck-mounted guns.

Comparison of contemporary fireboat photos with those of the *Willett* and *Duane* twins vividly illustrates Croker's recommendations for smaller wheelhouses with their better visibility and barebones deck layouts. Each boat was 132-feet-long, had a 28-foot beam, a 9-foot draft, weighed 222 tons and reached a speed of 15.38 statute miles per hour during trial runs.

But it was down in the engine rooms of the two nearly identical boats where the General Electric-built system of Curtis steam driven turbine centrifugal pumps excited the most interest while demonstrating their superiority over the old piston pumps.

Instead of steam boilers driving double cylinder, vertical inverted reciprocating piston pumps, the new look seemed overly simplistic. Mounted below deck in each boat were two turbine systems. Each turbine was rated at 600 horsepower and, via a horizontal shaft, drove a two-stage centrifugal pump.

During pump tests May 26, 1908, each boat delivered a maximum of 9625 gpm at 200 psi, with a vacuum of 27 inches and steam pressure gauged at the pump of 175 pounds, while the turbines were putting out 1900 revolutions per minute (rpm). When operated in series, the centrifugal pumps could put out 4500 gpm at 300 psi. To attain these higher water pressures the turbines were simply run at higher speeds.

Croker and others observing the trials readily noted the elimination of strain vibrations on the hull with the use of turbines while recalling that many an engineer and stoker wondered if the older boats would pop their rivets and welds during pedal-to-the-metal operations at major fires. Heightened safety was a fallout of turbine technology.

If the primary benefit of these new turbine systems was better steam production compared to the old boats and resulting increased water discharge and pressure, there was another significant value of turbines. The absence of valves and other moving parts would result in lower maintenance costs. Said Croker: "There is no doubt that the centrifugal pump turbine-driven, is far superior to the old style piston pumps."

With the FDNY's fireboat fleet standing at nine, largest in the United States, other city fire chiefs were reminded of the adage: "As the FDNY goes, so goes our fire department." No large city in the nation needed fireboats more than San Francisco, the busiest and most fire catastrophe prone port on the Pacific Coast. Since the 1848-1849 Gold Rush, severe fires periodically devastated the hastily-built city of wooden buildings clustered around brick commercial structures.

*Although New York laid claim in 1908 to the first fireboats with stationary steel gridwork towers with monitor guns, the **George H. Williams** of Portland, Ore. had the first elevated standpipe coupled to a nozzle four years before the tower-equipped New York fireboats were commissioned. The **Williams**, 105½-feet-long, favored a white color scheme in those days before red fireboats became traditional. The **Williams'** steam-powered piston pumps, soon to become outmoded by developing turbine technology, developed 6200 gpm. In service by May, 1904, the **Williams** answered 29 alarms during the next seven months. From the Collection of Bill Dahlquist*

Starting at 5:12 a.m. Wednesday, April 18, 1906, a cataclysmic earthquake and severe aftershocks touched off dozens of fires. Although San Francisco was surrounded on three sides by water, there was precious little in the quake-shattered mains to fight fires which, joining into a catastrophic firefront unequalled in American history, destroyed 28,000 buildings in an area of 514 city blocks, including the waterfront area.

*The **Governor Markham** was one of two tugboats which were pressed into service to supply water during the San Francisco Earthquake and Fire of 1906. The city had no fireboats. Pumping into a 3700-foot hose relay up Nob Hill, the **Markham's** efforts were futile and the Fairmont Hotel as well as other famous buildings in the area were lost. From the Collection of W.C. Dunn*

Firefighters used residual water in the mains, but without a fireboat, good supplies could not be drafted and relayed to steam fire engines. Given the scope of the disaster, no amount of fireboats could have saved the city. Two tugboats, the *Governor Markham* and its sister craft, the *Governor Irwin*, pumped into a 3700-foot-long hoseline relay up Nob Hill, but buildings in the area, including the Fairmont Hotel, were lost. Historical reports say at least 674 people were killed by the quake and fires, while damages exceeded $500 million. Years later, historians put the death toll at more than 3000.

The first major fire protection improvement after the earthquake and fire was the purchase and installation of an auxiliary water supply system consisting of two identical fireboats and eight sets of land-based salt water pumps utilizing turbine-centrifugal technology.

The fireboats *Dennis T. Sullivan*, named after the city's fire chief who was fatally injured in the quake, and the *David Scannell*, after another chief, were twin-stacked, 129-feet-long, with a 26-foot beam and a draft measuring 12-feet, 9-inches. Each mounted three guns: atop the pilothouse, above the deckhouse, and atop a structural steel tower near the stern.

The *Scannell* and *Sullivan* pumping systems were an advancement over those pioneered in New York. Instead of one steam turbine driving a centrifugal pump, the San Francisco boats had a General Electric turbine directly connected through flexible couplings to two multi-stage centrifugal pumps. The *Scannell* and *Sullivan* had two sets of these three-unit systems mounted upon a common bedplate.

While each boat was rated at 9000 gpm, each pump could deliver 2250 gpm at 150. The two pumps of each unit could, moreover, operate in parallel or in series. Pumping in parallel, they put out 4500 gpm at 150 psi. In series, pumping resulted in a delivery capability of 2250 gpm at 300 psi.

Evidence that San Francisco was determined to never again lack a good firefighting water supply was the system augmenting the fireboats. GE supplied eight land-based turbine systems capable of delivering 2700 gpm of salt water at 300 psi. Not counting San Francisco's fresh water resources for firefighting, the city's new system in 1908 consisted, in essence, of two 9000 gpm floating pumping stations and eight 2700 gpm land-based salt-water pumphouses. Altogether the system provided San Francisco with an additional 40,000 gpm fireflow.

The disasters in San Francisco, Baltimore, New York and Hoboken concerned Chicago fire officials, who knew their city was as vulnerable, perhaps more so, to a catastrophe along the heavily-industrialized south branch of the Chicago River. Two steam turbine-centrifugal pump fireboats were ordered to beef up the fleet. Only three fireboats were fully ready for major fires.

But the new boats would not arrive before a disastrous fire, which historical accounts agree would have been—except for the work of three fireboats and their crews—a repeat of The Great Fire of October 1871.

Acknowledged as Chicago's worst conflagration breeder was a heavily-congested cluster of grain elevators, warehouses, oil refineries and sprawling acres of railroad yards with hundreds of freight cars; many loaded with flammables. That district in the area of 16th and Canal Streets along the South Branch of the Chicago River was not only virtually inaccessible to land-based fire apparatus but lacking in water main supplies. The nearest hydrants were at least half-a-mile distant from the target hazard.

Fire Marshal James Horan warned building owners: "I do not believe that there is another place so bad in the city: no hydrants and no facilities for getting the engines to the scene of a fire." If the wind was right and the fire was bad, only fireboats could quickly get close enough to fight an effective battle. Considering their antiquated condition, even that hope was slim.

Weather conditions that Monday, August 3, 1908 were nearly identical to those immediately before Mrs. O'Leary's cow won immortal fame by kicking over a lantern which touched off the 1871 conflagration. It had been a long, dry summer as Chicago buildings preheated while fairly begging for an ignition spark, as hot winds swept across the city from the prairies to the southwest.

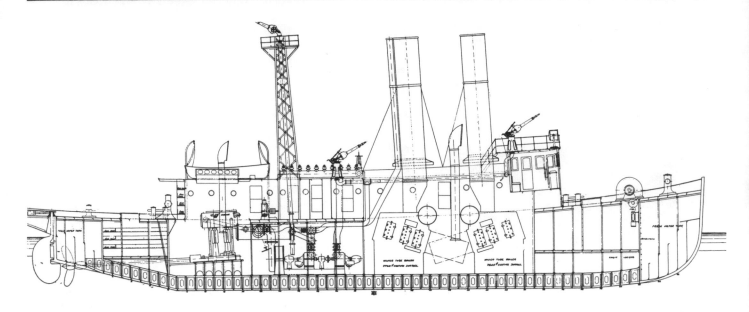

*Inboard profile of San Francisco's identical fireboats, the **Dennis T. Sullivan** and **David Scannell** which the city's Public Works System officially described as an auxiliary water supply system for fire protection and only incidentally as fireboats. Note the boiler configuration under the twin stacks and the steam turbine driven centrifugal pumps aft. Each boat was rated at 9000 gpm. Jim Delgado, National Park Service*

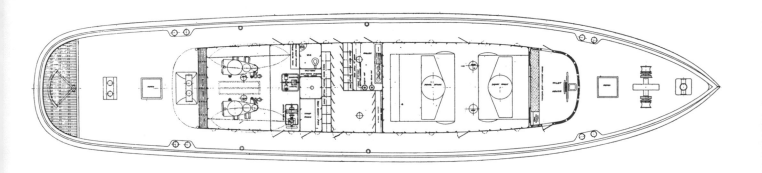

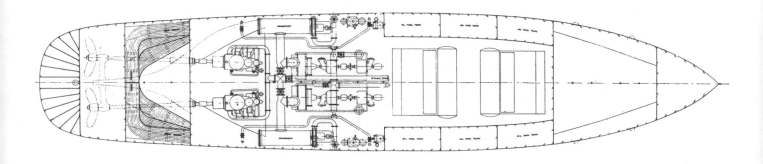

*Hold and deck plans for the **Dennis T. Sullivan** and **David Scannell** fireboats designed to augment San-Francisco's firefighting water supply system following the 1906 Earthquake and Fire which demonstrated the need for an improved water delivery system for firefighting in that city. Jim Delgado, National Park Service*

The twin-stacked **Dennis T. Sullivan** and the **David Scannell** each delivered 9000 gpm and had three monitor guns plus nozzles mounted along the bulwarks. Overall length of the boats was 125-feet with a length between perpendiculars of 120-feet, a moulded beam of 26-feet and a moulded depth of 12-feet, 9-inches. Jim Delgado National Park Service

San Francisco's **Dennis T. Sullivan** and **David Scannell** were among the first fireboats equipped with steam turbine driven centrifugal pumps. The identical boats each had two sets of these three-unit systems. Photo shows one of these sets with the two-multi-stage centrifugal pumps (left) driven by a General Electric Curtis steam turbine (right). From the Collection of Stephen G. Heaver Jr.

San Francisco's **David Scannell** under construction in 1909. From the Collection of Bill Dahlquist

The **Dennis T. Sullivan** shortly after its 1909 launching. The twin smokestacks of its sister fireboat, **David Scannell**, loom at the far right. From the Collection of Bill Dahlquist

Around 1 o'clock that afternoon, a worker in the Burlington Railway Company freighthouse near 16th and Canal flicked a lighted cigarette into the fully-packed freighthouse following the unloading of two Great Lakes freighters. Murphy's Law held true to form. What could go wrong did go wrong. The cigarette landed amid chemicals consigned to an explosives factory: soda, potash, saltpeter and nitroglycerine. A thunderous explosion erupted and totally engulfed the huge freighthouse that was bordered by two slips leading to the south branch of the river.

If that first explosion was felt for miles, those that followed were awesome, as flames swarmed into the adjacent Grain Elevators E and F, jointly-owned by the Burlington and Armour & Co. The 200-foot-tall elevators topped with wooden cupolas were filled with 800,000 bushels of wheat and corn. Even as fireboxes throughout the area were being pulled and workers fled, more explosions pocked the sky with flaming debris which set more fires, including wooden sidewalks.

Horan quickly arrived and sent in calls for more apparatus. "It was the hottest fire and the hardest to fight that I have seen in years," he said later. "When I arrived I could not get within 200 feet of the ( fire ).... I was in dread of a conflagration that would sweep all over the city. "

As additional apparatus arrived, Horan saw the horrendous task that lay ahead. Goaded by the wind, the blazing freighthouse and the two huge grain elevators were causing the fire to travel in a southerly direction. Flames would have to be stopped before they leaped the Chicago River and touched off more grain elevators, a huge warehouse and two oil refineries.

Inaccessibility by land companies, and a water shortage prevented an effective attack from the flanks. Only the fireboats could stop the fire where it was at its worst and most threatening. The *Illinois*, flagship of the fleet, made a stand in the slip between the freighthouse and the elevators and began battering the flames with its two bow turrets. The 10,000 gpm *Illinois*, 12-years-old and traditionally built with two coal-fueled Scotch double boilers, drove two sets of double duplex reciprocating piston pumps. Newest boat in the fleet, the 118-foot *Illinois*, also was the heaviest—287 tons—and had a 24-foot-beam and a draft of 11½-feet.

Also tied to a dock near the freighthouse and elevators was the 4000 gpm *Denis J. Swenie* (formerly the *Geyser*), 107-feet-long and mounting two Clapp & Jones piston pumps supplied by a Scotch boiler. Its single bow turret gun bored into the flames. The third boat, *Chicago*, was even stronger evidence of Chicago's need for bigger and better fireboats. Only 81-feet-long, the 3000 gpm boat had only one bow turret.

Despite its age—22 years—the *Chicago*, a former tug used for towing lumber barges, and its small pumping capacity, the fireboat crew held their station despite the radiant heat and omnipresent threat that those tall grain silos hulking above it that could, any moment, collapse. Without question, this was the finest moment in the history of the small but courageous *Chicago* and its crew, along with the two other boats which were pathetically ill-equipped to battle a fire of this magnitude.

Superhuman efforts and courageous feats were no less lacking on the land side. Sixty fire engines answered the 3-

11 alarm, plus three special calls for more help. Firefighters laid planks across railroad tracks and coaxed and tugged at the three-horse teams pulling the five-ton steam fire engines across the yards. They threaded their way around strings of smoking freight cars to get close to the fire. Ten fire horses fell over from heat exhaustion.

Despite the combined heat of the fire and hot afternoon sun, firefighters somehow managed to lay hoselines from pumpers taking suction at hydrants as far distant as a mile from the fire. Twenty-four engines made it to the river bank and began drafting water. Firefighters played cooling streams of water on those closest to the fire as the three fireboats stubbornly held their position while their turrets bored into the flames. These herculean efforts were not without more problems. Firefighters lugged heavy bags of coal more than half a mile to feed the insatiable boilers of the steam fire engines. Constant steam whistles from the fireboats signaled their need for more coal, too.

With the flames creating their own windstorms and the stiff breeze out of the southwest, tens of thousands of firebrands hurtled downwind and set 130 fires, mostly on roofs and awnings as far east as Michigan Avenue on the lakefront. Winds whipped the flames around and forced retreat from the railroad yards where 100 loaded freight cars burned to their metal wheels.

Still the worst of the onslaught was to the south as the storm of flames insisted upon bridging the Chicago River, despite the unyielding stand of the fireboats, whose crews worried that the wooden hulls and deckhouses would ignite from radiant heat, which raised paint blisters on the boats and cracked wheelhouse windows.

Spears of flame twice soared over the fireboats and touched off fires along docks on the opposite river bank. They were quickly doused. If the flames bested the fireboats, Armour's Union Elevator, filled with one million bushels of grain, would be lost, and so would the huge warehouse of Carson, Pirie, Scott & Co., the famous Chicago department store, as well as the N. K. Fairbanks and Standard Oil refineries.

Three antique fireboats shooting only 20,000 gallons of water a minute were pygmy-like under the towering flames and tall grain elevators. How they checkmated the southerly spread despite the heat and dense smoke that enveloped them remains a mystery, except for the obvious courage and stubborn refusal of the fireboat firefighters to surrender. Never in the history of the Chicago fireboat fleet would their bravery be surpassed.

But their efforts ultimately won the day. Around 3:30 that afternoon, Horan declared the fire was under control, although days of rugged firefighting lay ahead as damages soared to $1,508,000.

From the landside, the fire problem was diminishing. Not so on the river where the three fireboats continued to batter the gigantic mounds of burning grain that had spilled from the elevators. Pumping incessantly throughout the rest of the day and into the night, the three fireboats and their crews were close to dropping from exhaustion, heat and the harsh smoke.

At 3:30 the next morning, a shattering explosion thundered inside Elevator E. The towering cupola and the several hundred feet tall elevator wall fell outward and onto the *Illinois*. Pipeman Hans Hanson was knocked unconscious when a brick crunched his leather helmet. Others aboard the *Illinois* carried him off the boat as they fled onto the dock.

For a long minute, the *Illinois* refused to concede. The inch-thick ropes securing the boat to the dock tightened and snapped. The *Illinois* sank in 20 feet of water, while the two other boats steadfastly continued to pour water into the flaming ruins.

The battleline formed by those three old fireboats would long be a source of admiration among Chicago firefighters. "Notwithstanding the almost superhuman bravery and endurance of the men of the engine companies, the checking of the fire was largely due to the splendid work of the fireboats," says a contemporary history of the Chicago Fire Department.

"The *Denis J. Swenie, Illinois*, and *Chicago* poured thousands of tons of water on Elevators E and F and prevented the flames from spreading south," reported "A Synoptical History of the Chicago Fire Department," published in 1908 for the benefit of the widows and orphans of the Benevolent Association of the Paid Fire Department of Chicago. To which Horan added: "If the flames had got to the south we would have had another Great Chicago Fire."

As a postscript to the fire: the *Illinois* was raised from the river bottom, repaired and once again stood ready to fight fires, which it did for decades, until the boat was sold to a commercial tugboat company. When last heard from in 1960, the *Illinois*, renamed the *John Roen III*, was towing rafts of wood pulp logs across Lake Superior for Consolidated Papers, Inc., Wisconsin Rapids, WI. The *Chicago*, too old to fight modern fires, was retired and probably scrapped. The *Swenie* was condemned, towed out into Lake Michigan on December 29, 1927, and scuttled.

A year after Chicago's Great Grain Elevator Fire of 1908, the city modernized its fireboat fleet with two identical boats. Especially for proponents of steam turbine-centrifugal technology, the *Joseph Medill* and the *Graeme Stewart* were state of the art in electrical equipment. As was typical with fireboats of this genre, steam was supplied by two Scotch boilers to Curtis turbines, which operated the pumps.

The port and starboard mounted centrifugals delivered a minimum of 9000 gpm at 150 psi. In a tandem operational mode—acting as a single four-stage pumping system—delivery was rated at 4500 gpm at 300 psi. The twin-screw boats had variable speed reversing motors directly connected to the propeller shafts. Speed and directional control was from the pilothouse, thus obviating the necessity of the old-style engine order telegraph systems, although duplicate controls were located in the engine room for backup and safety considerations.

Just as Croker had said, the boats were highly maneuverable, thanks to the twin-screw concept. The pilot had the option of using the propellers as an aid to the rudders, while making short turns, or total steering by means of the propellers. Pilothouse control was by a system of electrical switches and rheostats.

A single-purpose generator supplied electrical current for the boats' above and below deck lighting, a fire service first. The boats also had a searchlight atop the wheelhouse for safer maneuvering in darkness and for providing frequently needed illumination at fires.

The *Medill*, named after the mayor elected soon after The Great Chicago Fire and who espoused fireboats, and the *Graeme Stewart*, named after another mayor, were each equipped with two turret guns: one near the bow and the other on a small raised platform amidship. Each turret was equipped with nine 3½-inch valves for hose couplings. The profile of both craft was typical of Great Lakes fireboats. Chicago's new boats could easily pass under the many bridges over the Chicago River without waiting for them to be raised.

Their specifications were fairly typical of fresh water fireboats: 120-feet-overall, a beam of 28-feet, a mean draft of 9½-feet, and displacing 500 tons. Each boat carried 2,800 feet of large diameter hose for supplying land companies, a common fireboat capability in Chicago as in most other cities.

The *Medill* and the *Stewart* were uniquely equipped with a system to hold them on station while they operated. Fireboats traditionally held their position while pumping by mooring to a dock or other fixed object. More often, however, boats were forced to operate immediately off shore. The backward thrust of the pumps' water pressure and flow at the nozzles pushed the boat away from the fire. To compensate for this problem, fireboats directed streams in a direction opposite from those battling the fire. This loss of one or more nozzles for firefighting curtailed the boats' firefighting capabilities and further resulted in a concomitant loss in water pumping and pressure capabilities for direct firefighting attack.

In an attempt to solve this problem which was as old as fireboats, the *Medill* and the *Stewart* had three telephone-pole-like spuds, or stabilizers. Two stood like stout masts towering port and starboard abaft the pilothouse and the third poked high near the stern. These spuds were similar to those used by dredge operators to hold their boats in place while working. The fireboats' spuds were of heavy steel tubing, 18-inches in diameter. Cast iron points were welded to the tips of the spuds to assist boring into the river bottom. The spuds were lowered from wells in the hulls by a system of steam-operated racks and pinions and could be extended to a depth of 24 feet.

Although the Chicago Fire Department is historically known for its apparatus innovations (crow's nests high up on water towers and the first fleet of elevating platforms or Snorkels, for example ) the spuds were duds. While there might have been some justification for spud-anchoring of the three fireboats that saved Chicago during The Great Grain Elevator Fire, these stabilizers were an unrealistic solution to an old problem. In the practical world of fireboat firefighting, boats need the capability to quickly change positions as fire conditions or emergency situations suddenly change.

The spuds defeated their purpose by severely restricting the quick maneuverabilty of the *Medill* and the *Graeme Stewart*, two of the boats' major features. There is evidence, moreover, that the hulking spuds made the fireboats top-heavy and that the rack-and-pinion system was not only cumbersome but prone to mechanical problems, especially during freezing weather. For these reasons fireboat history indicates other cities learned from Chicago's experience and no other fireboats were outfitted with spuds. Technology would, moreover, during the years ahead,

develop superior methods for holding fireboats on station.

Neither the *Medill* nor the *Stewart* were completely ready for service when delivered in 1909, but they were nevertheless called out when several grain elevators exploded and burned along the Chicago River. "The Marine Engineer" publication reported, "The Curtis-driven fireboats...gave excellent account of themselves in the fire which destroyed several grain elevators on April 29.

"The *Graeme Stewart* promptly responded to the first alarm at 4:30 a. m., and was shortly afterward in service with full pressure on the two gun nozzles. The *Joseph Medill*, although not in commission, went into action on a hurry call a few hours later with one of the gun nozzles and several hoses in operation. The operation of both boats was satisfactory in every respect," the magazine reported.

While Chicago, San Francisco and New York modernized their fireboat fleets, other cities preferred to stick with the tried-and-true old-fashioned boats with steam boilers and reciprocating piston pumps. Among them was a fireboat built by Crescent Shipyard, Elizabeth, N.J., for the Buffalo Fire Department. After an 18-day voyage via the St. Lawrence River and the old Welland Canal, the fireboat reached Buffalo November 6, 1900, "to the greatest reception ever given a boat in Buffalo" according to the logbook of the *W. S. Grattan*, named after a Buffalo fire commissioner.

The twin-stack *Grattan* was 118-feet-long, had a 24-foot beam and a 11½-foot depth of hold. With a top speed of 15 knots, the *Grattan's* coal-fueled boilers drove three piston pumps which delivered 9000 gpm. Its sturdy iron hull was ideal for extra duty as an ice-breaker to maintain shipping lanes and to service Buffalo's domestic water supply derived from the offshore Crib in Lake Erie.

A major reason why Buffalo bought the boat was the completion, two years earlier, of a fireboat pipeline: an underground gridwork of water mains supplying special hydrants to provide greater volumes of water to waterfront and downtown districts which were not only growing fast, but experiencing worsening fires. Typical of other cities with similar systems, the *Grattan* served as a floating pumping station to feed the fireboat pipeline.

Little did anyone foresee that this boat, which was old-fashioned even when it was new, would outlast turbine-centrifugal and gasoline-operated fireboats as it staked its unparalleled place in American fireboat history.

Just for starters: The boat, still operating as America neared the 21st Century, boasts the longest service life of any fireboat ever built. This, despite vigorous attempts time and time again to junk the boat. And just as many times, marine engineers who examined her declared she was not only seaworthy, but with a bit of facelifting would become a better fireboat at far less than the cost of a new one.

The *Grattan* went through name changes as it neared a century of service, was destroyed down to its Swedish iron hull during a fire, but like the Phoenix bird, rose from its ashes, was rebuilt and returned to fight still more fires. The *Grattan*—now the *Edward M. Cotter*—holds the unique distinction of being the only American fireboat to leave United States waters to fight a disastrous fire in a foreign city and to be cited for controlling the fire. But we get ahead of our story.

The **Illinois**, at 10,000 gpm, was the flagship of Chicago's fireboat fleet and won everlasting fame when it made its successful stand against flames during the Great Grain Elevator Fire of 1908. Debris from an exploding grain elevator silo fell onto the **Illinois** and sank it. From the Collection of Robert Freeman

Alone among American fireboats, Chicago's **Joseph Medill** and **Graeme Stewart** were outfitted with stabilizing spuds to anchor the boats to the river bottom while pumping. Note the spuds in their normal towering position abaft the wheelhouse and at the stern. To lock the boats into the river's muck the spuds were lowered as much as 24 feet. From the Collection of Steven Lang

Chicago's **Denis J. Swenie** was one of three antiquated fireboats which managed to prevent a sequel to the Great Chicago Fire when a horrendous industrial district fire in 1908 threatened the city. From the Collection of Robert Freeman

Steam-operated rack-and-pinion system (lower right) lowered 3 Chicago Fireboats' **Medill** and **Graeme Stewart** stabilizing spuds into the river bottom to hold the boats steady while pumping. The spuds were similarly raised to their normal position which caused them to seem like masts towering over the fireboats. Chicago's spuds proved to be duds. From the Collection of Stephen G. Heaver Jr.

The **Joseph Medill** was one of two fireboats Chicago acquired in 1909 to modernize its fleet. The **Medill** along with its twin, the **Graeme Stewart** could deliver 9000 gpm at 150 psi and represented the state of the art in steam turbine centrifugal pump technology. Unique to both boats were three stabilizer spuds—two abaft the wheelhouse on the port and starboard sides and the third near the stern. The spuds could be lowered 24 feet to the river bottom to hold the boats in position while pumping. From the Collection of Robert Freeman

Buffalo's **W. S. Grattan** began fighting fires in 1900 and stands ready to fight them today, despite almost total destruction by fire in 1928 and many attempts to junk her. The boat, renamed the **Edward M. Cotter,** has been in service longer than any other American firefighting apparatus, whether based on water or land. Buffalo Fire Historical Society

Boston's **John P. Dowd** was only two years old when called to battle this three-alarm fire in September, 1911, at the Warren Coal Co. wooden storage bunkers on Dorchester Avenue, South Boston. Also known as Engine 47, the **Dowd** was designed by William F. Keough and built by Bertelsen & Petersen at a cost of $92,625. Rated at 6000 gpm, the, **Dowd** was 113-feet-long with a 26-foot beam and a 9-foot draft; ideal for shallow water operations. The boat displaced 179 tons and served until 1941. From the Collection of Bill Noonan

The first fireboat of Portland, Me., **Engine 7**, followed conventional tugboat lines and mounted a turret gun at the bow and another atop the wheelhouse. Built in 1905, the boat was rated at 1650 gpm via its steam-powered piston pumps. From the Collection of Steven Lang

# CHAPTER FOUR
## THE ANATOMY OF A FIREBOAT

For a city built mainly of wood and whose economy in the early 1900's depended upon the Pacific Northwest lumber trade, Seattle's fireboat, *Duwamish*, was, except for its pilothouse, notably lacking in wood construction. This irony did not escape lumber merchants who would have liked to have supplied the construction materials for what everyone realized was going to be a prestigious boat. Pacific Coast shipbuilders, moreover, preferred wood because it was plentiful and cheap.

There was a strong rationale for the *Duwamish's* all-steel riveted construction which made it the sturdiest fireboat on the Pacific Coast, if not the entire United States. Stoutly-reinforced steel plating and iron supports were necessary if Seattle was to have a fireboat equipped to withstand the tremendous water volumes and pressures required to battle problems unlike those in any other American port.

Lumber supplier to the nation, Seattle was only 12 miles long, but had 75 miles of waterfront. Lining the shallow waters of Puget Sound were acres of lumberyards dotted with gigantic sawmills, wood-drying kilns and large wooden warehouses, often with wood vessels docked alongside them. Both high pressure and large volumes of water had to be lobbed hundreds of feet from the shoreline into burning areas stretching well back into the industrialized waterfront.

The *Duwamish* inherited a legacy of major waterfront fires, including the 1899 catastrophe that resulted in Seattle's first fireboat, the wood-hulled 5765 gpm *Snoqualmie*. But the *Snoqualmie* was aging and its water output was far below requirements of the booming waterfront, especially

*Seattle's **Duwamish** was unique among American fireboats because of its reinforced iron bow built to ram and sink burning lumber schooners if the **Duwamish** could not put out the fire. Stationed at the foot of Madison Street, its firehouse (left) was damaged during the Grand Trunk Pacific dock and warehouse blaze in 1914. The disaster was the first general alarm fire in Seattle's history and was the **Duwamish's** first major challenge. From the Collection of Bill Dahlquist*

*The coal-fired, twin smokestack **Duwamish**, commissioned in 1909, had a 9000 gpm capacity and mounted eight larger than normal guns. The **Duwamish's** ability to discharge large volumes of water at high pressures ideally suited the boat for protecting Seattle's waterfront lumber industries. From the Collection of Steven Lang*

*The **Duwamish** was a firefighting battleship whose pilothouse-mounted gun packed a 7000 gpm punch which could lob streams hundreds of feet beyond the shoreline while battling Seattle's lumberyard fires. From the Collection of Steven Lang*

*The pie-shaped pump order telegraph in the **Duwamish** pilothouse and duplicated in the engine room was typical of those in American fireboats. The pilot called for the desired pumping mode by turning the handle which moved the black arrow to the marked wedge and simultaneously rang an attention-getting bell in the engine room. Each wedge indicated an order: standby, start or stop pumps, increase or decrease water pressure. Jim Delgado, National Park Service*

after the 1897 Alaska gold rush made Seattle the gateway to what was to become our 49th state. Port traffic shortly would increase even faster with the 1914 opening of the Panama Canal.

Mindful of Seattle's shallow waters, Naval Architect Eugene L. McAllaster designed a boat without an external keel. The *Duwamish* would be 120-feet-long with a 28-foot beam and, perhaps in deference to New York Fire Chief Croker's generally adopted concept of deck simplicity, the *Duwamish* had a single flush deck of riveted steel plates. Although weighing 322 tons, the *Duwamish's* depth of hold was only 9½-feet; thus fulfilling specifications for a shallow draft boat.

The most prominent feature of the *Duwamish* was the configuration of the hull with its oversized and sharply-angled bow that projected like a battering ram at the waterline. This ramming bow, unique among fireboats, was of reinforced iron. Lumber schooners were mostly wooden-hulled. The purpose of the ram bow was simple: If the *Duwamish* could not control a fire aboard a burning schooner, the fireboat would ram and sink it in shallow water where the cargo, if not the vessel, could be salvaged. There is no record of any other fireboat built with a ramming bow.

The *Duwamish*, with no external keel and with a peculiar snout, was built at the Richmond Beach Shipbuilding Co. yard north of Seattle. Shunning the latest turbine-centrifugal technology preferred in New York, Chicago and San Francisco, McAllaster opted for the traditionally proven steam boilers and piston pumps. Four coal-fueled Mosher water tube boilers drove the twin-screw propellers so the boat could achieve 10½ knots; a top speed which surely was no record-breaker. On the other hand, the *Duwamish* was built not for speed, but as a firefighting battleship.

The boilers powered three American LaFrance double vertical steam piston pumps, each rated at 3000 gpm, for a maximum output of 9000 gpm at 200 psi. While most contemporary fireboat designers were mounting only several fixed monitor nozzles, McAllaster called for eight larger than normal guns. Three of them were on the port and three on the starboard sides. A seventh gun was aft and the eighth atop the pilothouse.

With that pilothouse gun the *Duwamish* could punch out thousands of gallons of water through an unusually long nozzle which had a 6-inch-diameter barrel and a 4½- inch tip. Its 7000 gpm wallop would become famous for knocking over burning boxcars loaded with lumber while the *Duwamish* fought lumberyard fires. Walls fell and telephone-pole-thick supports snapped like matchsticks under the onslaught of the gun. Including its manifold-supplied hoseline outlets, the *Duwamish* was claimed to be able to put out 24 streams, although this spectacle was usually reserved as an impressive welcoming spray for new ships or visiting dignitaries.

The *Duwamish*, its shiny black hull gleaming in the sunlight and its 45-foot twin smokestacks hulking over the boat, was commissioned in 1909 and became a major tourist attraction during that year's Alaska-Yukon-Pacific Exposition in the bustling city. Seattle's *Duwamish*, named after the river that coursed through the city, seemed ready for whatever the 20th century brought.

The *Duwamish* was berthed at the foot of Madison Street, beside the firehouse of Hose 5, a motorized rig. The location was strategically ideal, especially with construction in 1910 of the Grand Trunk Pacific Dock (Pier 53) adjacent to the berth of the *Duwamish*.

The Grand Trunk pier, warehouse and office building was the largest wooden structure on the Pacific Coast: 500-feet-long, 105-feet-wide and supported by no less than 5000 stout and highly-flammable creosoted wooden pilings. Towering above the dock was a clock tower and a three-story wooden warehouse filled with nearly 3 million board feet of lumber. Grand Trunk lacked a fire protective water sprinkler system and was virtually bereft of fire barriers.

If Grand Trunk officials and dockworkers worried about fire, they did not show it. Hose 5 firefighters became accustomed to answering an alarm a day for a smoldering fire, thought to be caused by a discarded cigar or pipe tobacco, falling between the joints of the wood planked deckwork.

Seattle firefighters often said that if the city was ever going to have its first general alarm fire, it would happen when luck finally ran out on the Grand Trunk. Many were lulled into a false sense of security with the mighty *Duwamish* berthed alongside the pier. Foresight said this was the best possible fire insurance. Hindsight would say otherwise.

Shortly after 3:30 Thursday afternoon, July 30, 1914, Grand Trunk workers saw black smoke puffing up alongside the dock where the freighters *Athlon* and *Admiral Farragut* were taking on cargo. A stevedore ran to the nearby Hose 5 firehouse and turned in the alarm at 3:46 p.m. Hose 5, driven by Fireman Patrick Cooper, 30, chugged to the fire and began the attack, but firemen were quickly disoriented by the rapid buildup of smoke and heat.

The crew of the *Duwamish* cast off and came around the end of the dock. Entering the slip, the fireboat was delayed by the *Athlon* and *Farragut* fleeing to open water. Additional alarms were sounded as the quickly propagating fire filled the warehouse with superheated smoke. The flashover that soon followed engulfed most of the structure and drove mountainous clouds of black smoke boiling so high in the sky that the loomup could be seen throughout Seattle and for miles around.

Fifteen minutes after the first alarm, Signal 6-6-6-6 was struck at 4:14 p.m. on fire alarm register bells in every city firehouse. This first general alarm in the history of the Seattle Fire Department called all 200 on-duty firefighters, plus 127 off-duty firemen, along with aerial ladders, hose wagons, 12 engines and the *Snoqualmie*.

The *Duwamish* churned down the smoke and heat-clogged slip and began battering the flames with its guns. As fire spread with incredible speed across the roof of the doomed Grand Trunk warehouse, radiant heat shimmered across the 167-foot-wide slip and touched off a major fire in the unsprinklered Colman Dock and its warehouse. Fire, smoke and heat were now on both sides of the *Duwamish*. Its steel hull became too hot to touch. On deck, the soles of the crews' rubber boots softened and melted.

Arriving about this time, the *Snoqualmie* began an attack in the opposite slip where radiant heat and smoke were bad. But the threatened Galbraith-Bacon Dock and warehouse was saved by its sprinkler system and the work of the *Snoqualmie*'s guns.

The boats rescued workers and Hose 5 firemen who had finally made it out of the inferno and jumped off the dock as flames whiplashed throughout the Grand Trunk complex. Hose 5's Driver Cooper and Fireman John W. Stokes, both badly burned, were pulled from the water. Jutting hundreds of feet out into the water, the Grand Trunk Dock was almost inaccessible to land-based companies, despite their excellent water supply from the special mains the city had the foresight to build in preparation for major fires.

With its heavy guns relentlessly pounding away at the flames, the *Duwamish* was forced to tie up to a ferry landing. "Owing to her shallow hull, she cannot hold a position with her streams going," explained Fire Chief Frank L. Stetson. "With a bow nozzle of 4½-inches and three 2-inch side nozzles going, there was not much difficulty in confining the fire to the Grand Trunk side."

Attacking on both sides of the slip—the *Snoqualmie* with her 5765 gpm capacity—and the *Duwamish*'s 9000 gpm, more than 15,000 gallons of water a minute pounded at a fire that refused to go out. The *Duwamish*'s battering ram streams knocked down walls and snapped pilings while slamming into the flames. The fire retaliated by damaging the *Duwamish* and Hose 5 firehouse. Already gone was Hose 5. The thick planking of the dock burned through and the apparatus, together with all 750 feet of its hose, dropped into the water, never to be recovered.

*Seattle's aging* **Duwamish** *required replacement of much interior plumbing. As rebuilt during the 1930s, a large-diameter manifold pipline, with valves controlling the water flow to guns, formed an exterior rectangle around the deck; a feature rarely, if ever, seen among American fireboats. Jim Delgado, National Park Service*

The unsprinklered Grand Trunk was built to burn and that's exactly what it did with losses, including those at adjacent structures of $773,952. Except for the guns of the *Duwamish* and the *Snoqualmie*, which together had 16 monitors shooting water at one time, the losses would have been worse. As many as six persons, including Hose 5's Cooper, who died three days later, were killed as the *Duwamish* got its first baptism under fire.

Uncounted lumber mill, shipyard, vessel and warehouse fires were in the *Duwamish's* future as a special mystique cocooned the fireboat which was to become an historic landmark on Seattle's waterfront. During the 1930s, marine engineers found that the manifold supply lines needed replacement. As rebuilt, the large-diameter manifold pipeline, with control valves, formed an exterior rectangle around the deck; a feature rarely, if ever, seen among American fireboats.

The *Duwamish* was 40-years-old in 1949 when its high operating costs and slow speed mandated a major overhaul. The boilers were replaced by surplus U.S. Navy diesel-electric engines. The one-of-a-kind ramming bow was reconfigured. With steel-hulled vessels in commercial trade, there was no need for ramming and sinking the long-gone wood hulled lumber schooners. Rebuilt, the *Duwamish's* profile became that of a conventional fireboat.

The old steam-driven piston pumps gave way to a pair of diesel electrically-driven De Laval Steam Turbine Company centrifugals which boosted the *Duwamish's* pumping

*After answering her last alarm in 1984, the 75-year-old **Duwamish** was retired to await her destiny, perhaps as a tourist attraction, while docked at Seattle's Lake Washington Ship Canal, Chittenden Locks. Jim Delgado, National Park Service*

*After 40 years of service, Seattle's **Duwamish** was rebuilt in 1949. The ramming bow was removed and the boat's pumping capacity was beefed up to 22,850 gpm at 150 psi. This gave the **Duwamish** claim to being the world's most powerful fireboat. From the Collection of Steven Lang*

capacity to 22,800 gpm at 150 psi to give the fireboat claim to being the world's most powerful. The boat also was outfitted with four underwater jet nozzles for better maneuverability and stability while battling fires. Radar and a depth finder were added, too.

The *Duwamish* helped to protect Seattle's waterfront for another 35 years until it answered its last alarm, September 8, 1984, when it raced to a fire on West Marginal Way. The continuing high cost of maintenance, the low availability of replacement parts and her age—75 years—and slow speed, ultimately forced retirement of the *Duwamish*. Other cities relegate their old fireboats to scrapyards, sell them as tug-

boats or sightseeing barges, or simply take them out into deep water and scuttle them.

Not Seattle. Still an historic landmark, the *Duwamish* patiently awaits her destiny while docked at Seattle's Lake Washington Ship Canal, Chittenden Locks. Many ideas have been proposed for her future, including a restaurant attraction and as a museum. By the end of 1988, it was not known what her ultimate use would be. But nobody dared suggest the possibility of condemning her to a graveyard for old fireboats. Whatever her future, the *Duwamish*, her boxcar-knocking-over monitor guns forever stilled—will somehow continue to make Seattle proud of its maritime heritage.

Less than three years after Seattle's Grand Trunk Pacific fire, the United States entered World War I; a point that has several bearings upon fireboat development. The nation was enjoying an economic boom at the start of the war and Pacific Coast ports were noting increased activity from Panama Canal traffic. San Diego was a notable example. The city was growing into a major port with naval installations.

Considering the war effort and increasing hazards along its waterfront, San Diego could no longer afford to be without a fireboat. The city did not lack boat building expertise, but it did lack readily-available coal for operating the boat. Southern California was distant from developed sources of coal. Even if there were mines, there was a national shortage due to the war.

So critical was this shortage and the need so great for fuel to power warships and freighters shuttling between the United States and Europe, that U.S. Fuel Administrator Harry A. Garfield ordered all eastern factories not making wargoods to close for two weeks. For San Diego, the old-style coal-fired boilers and piston pumps were as much out of the question as were steam-operated turbines.

An axiom in American fire service history says that there is nothing like an unmet need to bring out innovative solutions by firefighters. San Diego firemen decided to build a

*San Diego's **Bill Kettner**, built in 1918 by that city's firemen, was the first gasoline-operated fireboat in service on the Pacific Coast. At one time the 5000 gpm **Kettner** had a water tower, closely resembling a lifeguard station, immediately abaft the pilothouse. This July 2, 1959, photo was taken two years before the **Kettner** was decommissioned. The Firehouse Museum, San Diego*

fireboat themselves and to power it with gasoline which was abundant among Southern California's many oil producers, who were vigorously promoting its use.

Working in the San Diego Fire Department shops under the direction of Fire Chief Louis Algren, the firemen built the boat in 1918 and named it after a local congressman. The single-propeller *Bill Kettner* was 58-feet, 9-inches long, had an 18-foot beam, was of shallow draft and weighed 55 tons.

The *Kettner* was powered by three 220-horsepower Atlas-Imperial gasoline engines which enabled the boat to do 10 knots. Two of these engines had the dual function of driving the *Kettner*'s two 115-horsepower Seagrave two-stage centrifugal pumps which gave the boat a 5000 gpm capacity. The *Kettner* could simultaneously discharge 13 streams, including those from the three monitor guns which had a 360-foot reach.

Popular fire service history claims the *Kettner* was America's first gasoline-operated fireboat, but the claim must be shared with another. At least eight years before the Kettner, Marine Architect Arthur D. Stevens proposed to build a 4500 gpm gasoline-operated fireboat for Jacksonville, Fla. "The use of gasoline on so large a scale did not appeal to the fire department which was still wedded to the steam fire

engine and had no use for gasoline," said Stevens. The proposal was pigeon-holed.

After World War I, American fire departments were rapidly selling or pasturing their fire horses and auctioning their steam fire engines in favor of gasoline-operated apparatus. In 1918, Stevens' idea was revived when Jacksonville bought the war surplus submarine-chaser, *S.C. 145*, and hired him as engineer-in-charge of converting the 110-foot-long wooden sub-chaser into a fireboat. This marked the start of an era during which many cities cheaply acquired surplus Navy, Army and Coast Guard craft and rebuilt them as fireboats.

Following radical alterations, including the addition of 10 tons of iron ballast in the aft end of the engine room bilge to provide a more even trim for the heavy propulsion and pumping engines, the sub-chaser was rechristened the *John H. Callahan*. It was 110-feet-long with a 14-foot beam and an 8-foot depth of hull.

Propulsion powerplants were two Standard 220-horsepower, 6-cylinder air starting and reversing motors, each driving the *Callahan*'s two screws and their three-bladed propellers. Aft of each propeller was a bronze rudder. Forward of these motors were four three-stage De Laval centrif-

*After World War I and continuing for many years, fire departments often acquired surplus Army, Navy and Coast Guard craft and converted them into fireboats at costs far below those of building a fireboat from the keel up. Baltimore obtained this 1917 submarine-chaser, S.C. 428, from the Navy and turned it into a fireboat for* only $10,000. In service June 21, 1921, the 3000 gpm wood-hulled boat was 110-feet-long, had a 15-foot beam and a 6-foot draft. The **S.C. 428 (Cascade)** was berthed at the quarters of the historically famous **Cataract** (left). Charles Cornell

ugal pumps. Each was connected to a 300 horsepower Sterling 8-cylinder motor. Displacing 110 tons, the *Callahan* was rated at 7000 gpm and mounted two monitors: one atop the pilothouse and the other aft.

The *Callahan*'s fuel compartment had a capacity of 1200 gallons of gasoline. Except for Stevens' foresight, the fireboat could have been a floating gasoline bomb tucked below deck in a space of only 50-by-14 feet with 46 gasoline motor cylinders delivering a total of 1650 horsepower. With vapor leaks a prime consideration, Stevens provided a ventilating system of blowers, stacks, fans, exhausts and 16 large open air ports on the port and starboard hull adjacent to the pro-pulsion and pumping equipment. The $50,000 conversion was a fraction of the cost of a fireboat built from the keel up. This factor caused cities to scout war surplus boat lists for craft they could convert into fireboats.

That coal-operated fireboats were becoming an anach-ronism was dramatically shown by Stevens' comment: "The engines at 1000-1500 rpm run very smoothly ... with a con-sumption of less than 0.66 pounds of gasoline per horse-power hour." Gasoline-operated fireboats could be operated with less manpower, another cost saving. Unlike steam boil-ers which consumed coal around-the-clock while standing

ready for alarms, gasoline motors were more economical because they merely had to be started only when the fire-boat was needed.

Even staunch supporters of tried-and-true coal use were impressed. They could recall tests of Chicago's steam-driven turbine-centrifugals, *Joseph Medill* and *Graeme Stewart*. Their tests showed they burned nearly 1½-tons of coal an hour or 13½ tons during eight hours of steady pumping.

The *Callahan*'s maneuverability was similarly impressive. "At full speed (12 knots) the boat was brought to a complete stop in less than 1½-lengths distance and 22 seconds time," said Stevens. "A complete turn was made at full speed in a 125-yard-diameter circle. The boat also answered the helm while backing with both engines."

Communications between shore and fireboats, a problem as old as the history of fireboats, also saw postwar innova-tions. Boston, for example, had used flashing red lights atop fire headquarters to tell fireboat crews whether they were to continue to the fire or return to their berths. Sema-phores, megaphones, flags and lantern signals gave way to wireless telegraph.

*Marking the beginning of the end of steam fireboats was the **John Purroy Mitchell**, built in 1921. This last steam fireboat built for New York utilized steam turbine technology. The 9000 gpm at 150 psi fireboat ended another era: It replaced **The New Yorker** as the flagship of the FDNY's Marine Division. Built for $275,000 at Standard Shipbuilding, Shooters Island, N.Y., the steel-hulled* *Mitchell was 132-feet-long, had a 27-foot breadth, an 11-foot draft and weighed 334 tons. Named after a New York mayor, the five-gun Mitchell served until February, 1966. The fireboat (right background) probably is New York's **William J. Gaynor**. From the Collection of Bill Noonan*

New York experimented with telegraphy between Manhattan fire alarm headquarters and the *James Duane*. It seemed like an ideal solution to communications problems but the city officials decided it was too costly to pay telegraphers around the clock when their service was so seldom needed. The Boston Fire Department was the first to install a two-way radio communications system in the American fire service, starting in October, 1923. Initially, two-ways were mounted on fireboats and later put aboard land-based apparatus. The idea soon was adopted by other cities.

With gasoline-operated fireboats usurping coal-fired piston and steam turbine technologies, it was but a brief time before the advent of dieselization. Houston's *Port Houston*, built in 1925, was the first fireboat utilizing diesel generators. The 8000 gpm *Houston* was powered by three Winton diesels which developed 1165 horsepower. These were connected to electric generators for both propulsion and pumping. The *Houston* was built by Bethlehem Steel at its Wilmington, De., yard from designs by Cox and Stevens. The boat was 118-feet-long and had a beam of 27-feet.

But dieselization was many years away from becoming universally desired as the power supply of first choice. Many fireboats were converted to diesels only after their aging equipment and high maintenance costs forced them into drydock for major renovations. Until the crossover to diesels, new state of the art fireboats continued to be gasoline-powered, including *Los Angeles City No. 2* (later the *Ralph J. Scott*).

Considering the high costs of construction, maintenance and the number of crew members necessary for a firefighting apparatus that is seldom used, fire chiefs historically have found that fireboat funding is difficult to justify to city officials. Fireboats are, however, like insurance policies. You hope you never need to make claims upon them, but appreciate their value when their costs are more than justified by only one major waterfront fire.

The genealogy of *Los Angeles City No. 2* is a textbook example of the anatomy of a fireboat. The city acquired a

America's longest fireboat is the Port of New Orleans Dock Board's **Deluge**. The steel boat is 138.8-feet long, has a 29-foot beam, a 14.6 foot draft and displaces 370 tons. Steam powered when it went in service in 1923 as shown here, the **Deluge** was later dieselized. With a 10,400 gpm pumping capacity, the **Deluge** has four monitors: one on the pilothouse, two on the second deck and a gun mounted on a tower 35 feet above the main deck. Almost unique among large American fireboats, the **Deluge** does not have a bow monitor. Additional water delivery is provided by 10 portable, rail-attached monitors supplied by 3-inch hose. The single propeller **Deluge** is berthed at Algiers on the Mississippi River across from New Orleans' famous French Quarter and is rated at 15 knots. Consistent with the needs of the port, third largest in the United States, the **Deluge** is capable of producing 20,000 gallons of foam and discharging it at the rate of 600 gpm. The foam is ideally discharged via the pilothouse monitor. The **Deluge** also serves as a port tugboat and its powerful water discharging capability is used to wash away silt along the Mississippi's banks. From the Collection of Paul Ditzel

harbor with the August 28, 1909, annexation of San Pedro, Wilmington and East San Pedro (Terminal Island). The LAFD immediately proposed the purchase of a fireboat to protect the rapidly-growing port, but had to compromise with the lease of two tugs with firefighting capabilities on an as-needed basis.

Los Angeles' first fireboat, **Aeolian**, probably was one-of-a-kind because its sole firefighting capability was a 60 gallon bicarbonate of soda and sulphuric acid chemical supply which, when mixed and discharged through the reeled hose, formed an extinguishing agent. Purchased in 1915 and believed to have been built in Seattle, the 20-foot-long craft went in service in 1916. From the Collection of Bill Dahlquist

The department persisted. In 1915 Los Angeles bought its first fire boat, *Aeolian*, which was still another compromise. The *Aeolian* was little more than a 20-foot-long craft mounting a 60 gallon tank of bicarbonate of soda and sulphuric acid which, when mixed, formed a fire extinguishing agent. The boat also carried a short ladder for boarding lumber schooners and other vessels. Augmenting the *Aeolian* were two steam fire engines that were loaded onto a barge and towed to fires.

Ten years after annexation, Los Angeles still did not have adequate protection for its eight miles of waterfront, despite increasing harbor traffic, including that from the opening of the Panama Canal, plus a burgeoning petrochemical industry. Los Angeles, moreover, became the world's largest importer of lumber to meet construction requirements attendant upon the city's phenomenal growth.

The fire department, doggedly lobbying for a fireboat of 9000 gpm capacity, had to settle for a city council decision which, in 1919, gave Los Angeles its first true fireboat: the 65-foot-long *Fireboat No. 1*, which could only deliver 2000

In 1919, Los Angeles acquired its first true fireboat, but it could deliver only 2000 gpm, not the 9000 gpm capacity the fire department had hoped to buy. A pilothouse was later added to **Fireboat No. 1**. From the Collection of Bill Dahlquist

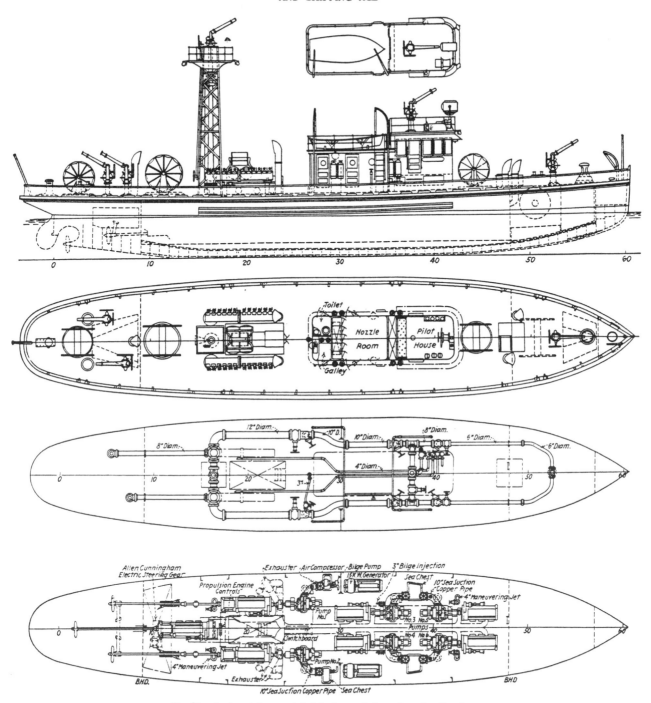

Profile, deck, piping and hold plans of Los Angeles fire boat

*L.E. Caverly drew plans in 1925 for the state of the art **Los Angeles City No. 2** with Fire Chief Ralph J. Scott's input. The 99-foot-long fireboat was designed with seven gasoline-powered Winton engines and four Byron Jackson all-bronze centrifugal pumps to provide a 10,200 gpm capacity. **Los Angeles City No. 2** had many unique features, including a water tower gun that could be extended, a 10,200 gpm Big Bertha cannon, three propellers, a craft that was at that time the fastest American fireboat, a large capacity of chemical foam for petrochemical fires and a sprinkler system to protect the hull and deckhouse. From the Collection of Paul Ditzel*

*Los Angeles City No. 2*, a state of the art fireboat with many innovations, was built in 1925 at the Los Angeles Shipbuilding and Drydock Corp. (Todd Shipyard), San Pedro. The $214,000 fireboat, later renamed *Fireboat 2*, the *Ralph J. Scott*, originally was gasoline-powered and rated at 10,200 gpm. From the Collection of Bill Dahlquist

Mounting 5 monitor guns, including a tower gun which could be extended 44 feet above water level, **Los Angeles City No. 2** was one of the first large fireboats powered by gasoline. Carrying 2156 gallons of fuel, the fireboat featured a safety system which completely changed the air in the engine room every five minutes as a precaution against leaking gasoline vapors. A further safeguard against below deck fires was a bank of 18 carbon dioxide extinguishing agent cylinders forward of the water tower. From the Collection of Paul Ditzel

Big Bertha, mounted atop the pilothouse of **Los Angeles City No. 2** put out 10,200 gpm at 200 psi and shot a stream 475 feet. From the Collection of Paul Ditzel

Seven 350-horsepower, 6-cylinder in-line Winton gasoline engine powered **Los Angeles City No. 2**. Three of these Wintons drove the center, port and starboard propellers for a top speed of 17 knots. The four other Wintons operated the forward-mounted pumps. Increased pumping capacity resulted from the dual capability of the two wing propulsion engines when they were switched from propulsion to pumping mode. From the Collection of Bill Dahlquist

gpm. That year, Ralph J. Scott was appointed chief engineer of the LAFD. Scott, who was to become world-renowned for his fire service innovations, immediately began a five-year push for adequate waterfront protection. On May 6, 1924, voters approved a $400,000 bond issue for construction of a fireboat, a station to house it and auxiliary apparatus.

Following completion of L.E. Caverly's designs dated January 31, 1925, a $214,000 contract was awarded to the Los Angeles Shipbuilding and Drydock Corp. (Todd Shipyard) in San Pedro. The contract called for an October 1, 1925, delivery date. At 10:15 a.m., October 25, 1925, Mrs. Ralph J. Scott christened *Los Angeles City No. 2* by breaking a bottle of fire-extinguishing foamite across the bow.

The white-painted fireboat was of riveted steel construction, 99-feet-long with a beam of 19-feet and a highly-desirable depth of hold of 9-feet, 7-inches. It weighed 152 tons and carried 2156 gallons of fuel. The powering system included seven aft-mounted 350 horsepower 6-cylinder in-line Winton gasoline engines. Three of them drove the center, port and starboard wing propellers for a top-rated speed of 17 knots. *Los Angeles City No. 2* was not only the fastest fireboat afloat, but the first to be built with three propellers.

The four-other Wintons drove the forward-mounted pumps. A feature of the propulsion powerplant was the dual capability of the two Wintons driving the wing propellers. They could be switched from propulsion to pumping mode, while the Winton mounted furthest aft was solely used for propulsion. Two 25-horsepower engines were connected to a pair of 15 kilowatt generators. One provided power for steering and lighting. The other was in reserve.

*Los Angeles City No. 2's* six pumps were Byron Jackson four-stage centrifugals. Mounted in pairs forward of the propulsion system, each of these all-bronze pumps was rated at 1700 gpm at 200 psi, for a total output of 10,200 gpm. A manifold system supplied water to five monitor guns and 24 hose outlets equally distributed on the port and starboard sides and were designed for coupling to standard 3½-inch diameter boat hose. Additional hose was carried on four pivoting reels, two fore above and below deck and the others aft. Clamps on port and starboard bulwark rails permitted attachment of nozzles—which came to be called rail standees—that were supplied by hose coupled to the manifold system.

Monitors were near the bow and stern. A third was mounted atop a 33-foot steel gridwork tower. This stand-

*Six of these all-bronze Byron Jackson four-stage centrifugal pumps were mounted in pairs forward of* **Los Angeles No. 2's** *Winton engines. Each pump was rated at 1700 gpm at 200 psi for a total output of 10,200 gpm. A manifold system (upper right corner) distributed water to five monitor guns and 24 hose outlets. Water was obtained from grated openings on the port and starboard sides of the hull which always filled the seachests. From the Collection of Bill Dahlquist*

*Mounting a fireboat-sized water cannon, this 1925 Mack hose wagon was stationed with **Los Angeles City No. 2**. Boat Tender 2 carried a large supply of 3½-inch diameter hose for fighting fires too distant from the shoreline for fireboat streams to reach. **Los Angeles City No. 2** not only supplied the Mack's cannon, but pumped into the boat tender's hoseline. From the Collection of Mort Schuman*

***Los Angeles City No. 2*** *battled its first major fire, March 3, 1926, when the fully-loaded lumber schooner, **Sierra**, burned at the E.K. Lumber Company wharf, San Pedro. From the Collection of Paul Ditzel*

pipe-supplied gun could be elevated electrically 11 more feet for a total rise of 44-feet above water level. This feature was a fireboat first and would be adopted by other fireboat designers.

Another striking feature of *Los Angeles City No. 2* was the water cannon mounted atop the pilothouse. With up to a 6-inch diameter tip, the gun was referred to as Big Bertha, probably after the famous German cannons which were mounted on railway flatcars during World War I. Big Bertha put out 10,200 gpm at 200 psi and shot a stream 475 feet.

Sensitive to the hazards of gasoline operation, Co-Designer Scott included a system of cowls, louvers, exhaust blowers and discharge ventilators which resulted in a complete air change in the engine room every five minutes. Further protection was provided by a bank of 18 carbon dioxide extinguishing cylinders, each containing 50 pounds of agent. They were located immediately forward of the water tower.

Other unique amenities included a perforated pipeline extending around the top of the deckhouse and another under the upper fender along the hull. This water-showering system would help to protect the boat from radiant heat. *Los Angeles City No. 2* was the first fireboat to be outfitted with a large supply of chemical firefighting agent: 300 gallons of foamite for petrochemical firefighting. When mixed with water, this agent expanded into 3000 gallons.

The boat likely was the first to carry smoke masks which were crude predecessors of oxygen-supplied breathing apparatus in standard fire service usage many years later. *Los Angeles City No. 2* probably carried them long before land-based apparatus did because of the highly-injurious creosote-permeated smoke commonly associated with wharf and pier fires. Land-based firefighters were not as constantly exposed to this smoke because they could be regularly rotated. When a fireboat gets on station it might be there for hours without relief.

*Los Angeles City No. 2* was commissioned December 2, 1925, and went in service 25 days later with a crew of 14, including a captain, a pilot, a mate, two engineers and fire-fighters, probably based with *Fireboat No. 1* at the foot of First Street on the Main Channel. Shortly thereafter, *Los Angeles City No. 2* moved into its new house at Berth 226-227, Terminal Island. The location enabled the boat to quickly reach any fire in the port. The station, one of few covered boathouses ever built for American fireboats, became a waterfront landmark until its demolition, July 22, 1986, to make way for a new cargo container complex.

Stationed in the firehouse was a new 1925 Mack hose wagon, Boat Tender 2, as well as foam-carrying apparatus. Since the horse-drawn era, fireboat tenders have often been stationed with or near fireboats. Boat Tender 2 carried 3½-inch-diameter boat hose into which the fireboat could pump while fighting fires too distant from the shoreline for fireboat streams to reach. The Mack also mounted a large water cannon which was also supplied by *Los Angeles City No. 2*.

Any discussion of fireboat anatomy must include constant maintenance; a problem worse in salt water ports where the corrosive effects are greater. With age and changing port protection needs, fireboats require facelifts. *Los Angeles City No. 2* got its first starting in 1945. The seven Winton propulsion engines were replaced by five 275 horse-

power, 6-cylinder, in-line Hall Scotts and two 625 horse-power V-12 Hall Scotts. Radar was added in 1956 and the boat, most often referred to by then as *Fireboat 2* was renamed the *Ralph J. Scott* on May 8, 1965.

Even with these remodelings, the *Scott* was 43-years-old in 1968, when Fire Chief Raymond M. Hill decided to sell the boat for its salvage value. Despite the affection and care that had been given the *Scott*, Hill had good reasons for sharpening his cost-cutting pencil. Maintenance costs were high as were those of paying 14 crewmen each 24-hour shift to keep the boat ready for a rare need.

Port protection problems were changing with large cargo container vessels replacing the old break-bulk freighters. Huge petrochemical installations and tankships demanded protection capabilities that the aging *Scott* could not meet. Los Angeles not only needed a boat with a beefed up pumping capacity, but one that was more efficiently operated, more maneuverable, safer, easier and cheaper to maintain and crewed with less manpower.

Meeting at the *Scott's* station, Hill and staff chiefs were about to rubber stamp his plan to get rid of the fireboat when Capt. Warner L. Lawrence, the junior officer at the conference, spoke up. Lawrence was not much of a pencil-pusher, but he did know waterfront firefighting. For 28 of his 40 years on the LAFD he had skippered the *Scott*. Nobody better appreciated the *Scott's* pluses, minuses and idiosyncracies.

Lawrence explained his plan to save the *Scott*. Hill, never known for his eagerness to change his mind, nevertheless listened. The *Scott's* hull was seaworthy and gave every indication of remaining so for many years as did the bronze Byron Jackson pumps. For safety and economy, the *Scott* should be dieselized. The seven gasoline engines would be replaced by three 380 horsepower, 6-cylinder in-line Cummins, two 525 horsepower V-12 2-stroke Detroits and and two 700 horsepower V-12 Cummins. The *Scott's* pumping capacity could, therefore, be substantially increased.

Lawrence suggested the replacement of the five hand-wheel monitors with 3000 gpm hydraulically-operated stainless steel guns. Remove the rail standee clamps, said Lawrence, cut holes in the bulwarks and mount six stainless steel nozzles, each capable of putting out 1000 gpm. Three would go on the port side and the others starboard: all toward the bow.

For the perennial problem of delivering efficient streams to batter hard-to-reach fires under wharfs and piers, Lawrence's plan called for a port and starboard 3000 gpm underwharf nozzle at the waterline near the bow. Superior maneuverability would be aided by four 2000 gpm underwater thruster nozzles with 2½-inch tips on the fore and aft port and starboard sides of the hull. These would enable the *Scott* to better hold its station while battling a fire and give it the ability to move sideways, port or starboard.

Lawrence conceptualized a remote control console to replace the old one in the pilothouse, thus providing the Scott with an electrically-actuated steering system. Below deck, a new Quiet Room, amidship and made of three-quarter-inch tempered glass, would muffle some of the deafening problems of propulsion and pumping noise. To meet the current and future needs of the port, Lawrence proposed that a power-operated lift boom with a basket at the end could raise firefighters and equipment 31 feet above the waterline and onto the decks of ships.

Chief Hill was impressed; the more so when he learned that renovation would cost $239,000, a fraction of that for a new boat. The rebuilt *Scott* could, moreover, be operated with eight, not 14 crewmen, for an annual savings of at least $100,000. Labor savings alone would quickly offset the cost of upgrading. As incredible would be the fact that during the years required for this major overhaul, the *Scott* would, except for short periods, remain in service.

Lawrence's plan saved the *Scott* which went into the yard of the Fellows and Stewart Division of Harbor Boat Building Co., San Pedro, for the job. On October 29, 1969, the boat returned to its station. With the gradual replacement of the gasoline engines with diesels by mid-1978, the stack of 18 carbon dioxide cylinders was removed.

Saturday, December 6, 1975, was the highlight of Lawrence's career as the *Scott*'s 50th birthday was feted during a harbor celebration that attracted thousands, including Mrs. Adeline Scott, widow of the fire chief. The 13-gun *Scott* emerged from its cavernous boathouse looking better than she did the day she was delivered.

Instead of her original white paint which had been changed to camouflage gray during World War II and the Korean War, the *Scott* sported a red hull, white bulwarks, deckhouse and tower. The five stainless steel deck and tower nozzles and a new Big Bertha glistened in the Saturday midday sun. Its deck was bright brick-red and the *Scott*'s huge hose reels were outfitted with snappy red jackets.

With Lawrence in the pilothouse, the *Scott* churned into the Main Channel, several hundred yards off the boathouse and opened up its guns in a water salute Los Angeles firefighters call a flower pot, because that is what the spray resembles. The rebuilt 3774 horsepower *Scott* was now rated at 18,655 gpm at 150 psi. As it pivoted in the Main Channel, the guns of the *Scott* registered even more: a discharge of nearly 20,000 gpm. The boat tilted slightly from the back pressure and the sunlight created a rainbow of colors as the *Scott* showed it was ready for many more years of service. Observing from the pilothouse from another fireboat, Hill nodded his approval. Lawrence, ever taciturn, said nothing. He didn't have to.

*Rebuilt Los Angeles **Fireboat 2**, the **Ralph J. Scott**, emerges from its cavernous boathouse following a $239,000 facelift. Notable among the changes were all new stainless steel guns, including six port and starboard bulwark monitors. Also new was a port and starboard 3000 gpm underwharf nozzle mounted abaft the bow near the waterline. Underwater port and starboard thruster nozzles mounted in the hull provided better maneuverability. Underwharf and thruster nozzles are remotely controlled from the reconfigured pilothouse with its new console. From the Collection of Paul Ditzel*

*The **Ralph J. Scott**, resplendent in its new outfit: red hull, white bulwarks, deckhouse, tower and brick-red deck, an elevating boom amidship and new stainless steel guns poses proudly in the Main Channel of the Port of Los Angeles. Its old gasoline engines were replaced and the 3774 horsepower dieselized **Scott**'s pumping capacity was increased from 10,200 gpm to 18,655 gpm at 150 psi. From the Collection of Paul Ditzel*

*Water cascades from three stainless steel 1000 gpm bulwark nozzles while 2000 gpm more shoots from underwharf nozzle close to the **Scott**'s water line. The 3000 gpm stream at the right is from the stainless steel bow turret. Hal Wakeman Photo From the Collection of Bill Dahlquist*

Fire Chief David Campbell of Portland, Ore., was killed, June 26, 1911. following explosions and a roof collapse during the Union Oil distributing plant fire. The city's new fireboat was named after the popular chief when it went in service the following year. Designed by Fred A. Ballin and built at the Smith and Watson Iron Works, the **David Campbell** was 85-feet long. Its original 9200 gpm capacity was later increased to 12,500. The **Campbell** served for 14 years. From the Collection of Bill Dahlquist

Philadelphia's **J. Hampton Moore** was built in 1921 at the Merchant Ship Corporation, Chester, Pa. The fireboat, virtually identical to the city's **Rudolph Blankenburg**, purchased at the same time, had a 10,000 gpm capacity. Displacing 292 tons, the **Moore** was just under 130-feet-long with a 28-foot beam and a 9-foot draft. Its firefighting capabilities included four deck gun; one mounted on a 28-foot tower. The boat served until 1950. From the Collection of Paul Ditzel

*The **Ralph J. Scott's** 50th birthday was celebrated, December 6, 1975, as the Los Angeles fireboat opened up its guns in a "flower pot" shaped water salute to thousands attending the waterfront celebration. With its veteran captain, Warner L. Lawrence, in the pilothouse, the boat pivoted in the Main Channel as the **Scott** registered nearly 20,000 gpm discharge. Sunlight reflecting through the Scott's spray created a rainbow as the fireboat showed it was ready for at least another half century of service. From the Collection of Paul Ditzel*

*Nine guns of three Boston fireboats attack a wooden structure fire. Three forward guns of the **Angus J. McDonald** (right of center) lob streams while those of the **Thomas A. Ring** (right) and the **John P. Dowd** (left) hit the flanks. The 3000 gpm **Ring,** in service starting in 1911, was the newest fireboat in the Boston fleet at the time of this fire. From the Collection of Bill Noonan*

# CHAPTER FIVE
## THE DAY THE *GRATTAN* BLEW UP

If Los Angeles' *Ralph J. Scott* was the grandfather of many fireboats because so many followed its prototype, then Seattle's *Alki* was the grandmother. "She is the Queen Victoria of the fireboat fleet, due to all her offspring," said the Seattle Fire Department, referring to the British monarch who reigned from 1819-1901 and had nine children.

The *Alki*, named after Alki Point where the city's first settlers landed in 1851, was designed by W.C. Nickum & Sons Company, Inc., predecessor of the Elliott Bay Design Group, Seattle, along with Chief Engineer George Mantor of the Seattle Fire Department. Like the city's *Duwamish*, the *Alki* was designed as a firefighting battleship. As built in 1927 in Oakland, the boat's 12,000 gpm capacity was one-third more than the *Duwamish's* original output. Built of steel and at first gasoline-powered, the three-propeller *Alki* is 123½-feet-long with a beam of 28-feet, a draft of 9½-feet and weighs 197 tons.

The *Alki* is armed with nine large monitor guns; three on the port and starboard sides amidship, one near the stern and another on the pilothouse. The ninth gun is tower-mounted on a boxy gridwork of steel extending 32 feet above the waterline. This gun can be extended an additional 10 feet by means of a telescoping tower similar to the earlier-built *Scott*. Said Chief Mantor at the time of the *Alki's* commissioning: "Try to visualize a stream 5-inches in diameter being forced through a nozzle at 90 psi and thrown a distance of over 400 feet and you have a conception of what the *Alki's* power really is."

The *Alki's* powerplant originally consisted of seven 1200-rpm, 300 horsepower Winton 6-cylinder gasoline engines. Three of these engines drove the center and the port and starboard propellers. The four other Wintons powered six Byron Jackson 4-stage centrifugal pumps. Fuel was supplied by a 5500 gallon capacity gasoline bunker.

Byron Jackson, a Berkeley, Ca. pump manufacturer, had developed pumps especially for the fire service. "Centrifugal pumps are superior to positive displacement (piston) pumps," said a company publication. "They are free from vibration, can idle at full pressure against a closed valve without damage to the discharge hose, give non-pulsating flow, require less space, (and with) less maintenance, are lighter (and) cheaper in first cost and are superior in ruggedness and dependableness." Proof of those claims is the longevity of Byron Jackson pumps which remained in use as dieselization became commonplace.

The pump maker further said, "Greater horsepower is obtained at remarkable saving in space and weight, enabling the installation of a multiplicity of pumping units in a hull of reduced size, resulting in vastly increased pumping flexibility and speed. The multiplicity of pumping units adds to the factor of safety involved, since the failure of one or two units still leaves greater capacity available than in a steam (fireboat) of even larger displacement."

Rebuilt and dieselized in 1947, the *Alki's* pumping capacity was boosted to 16,200 gpm at 120 psi. It has eight Byron Jackson pumps. Six are four-stage centrifugals driven by General Motors twin 6-71 diesels at 380 horsepower. The others are single-stage centrifugals powered by 20 horsepower Westinghouse direct current motors. The *Alki's* two 45 horsepower generators were built by General Motors.

*Seattle's **Alki**, launched in 1927, was designed by W.C. Nickum & Sons Company Inc., predecessor of the Elliott Bay Design Group, Seattle. Built in Oakland, Ca., the boat's 12,000 gpm capacity was later increased to 16,200. The **Alki**, named after the landing point of Seattle's first settlers, marked the start, along with Los Angeles' **Ralph J. Scott**, of a new breed of gasoline-powered fireboats with large pumping capacities. From the Collection of the Elliott Bay Design Group*

*Even with this 1918 drydocking renovation of Buffalo's **W.S. Grattan**, the fireboat was ill-equipped to fight a 1928 oil barge fire which resulted in an explosion as flames overwhelmed the boat. The **Grattan's** engineer was killed and four of her crewmen were severely burned. The boat's Swedish iron hull remained intact, however, and the **Grattan** was rebuilt. Buffalo Fire Historical Society*

# Fire Boats

Practically all cities having any large measure of water front development provide fire boats to lessen the fire hazard for ships, warehouses, and contiguous water front property.

For fire boat service, centrifugal pumps are superior to positive displacement pumps in every way. They are free from vibration, can idle at full pressure against a closed valve without damage to the discharge hose, give non-pulsating flow, require less space, require less maintenance, are lighter, are cheaper in first cost, and are superior in ruggedness and dependableness.

It is interesting to note that the latest fire boats have been equipped with gasoline engine driven centrifugal pumps. These modern fire boats have proven to be as reliable as the steam boats which they supersede. Greater horse power is obtained at remarkable saving in space and weight, enabling the installation of a multiplicity of pumping units in a hull of reduced size, resulting in vastly increased pumping flexibility and speed. The multiplicity of pumping units adds to the factor of safety involved, since the failure of one or two units still leaves greater capacity available than in the steam boat of even larger displacement.

This Company has developed a special pump for fire boat service. Our Engineering Department will gladly confer with naval architects or chief engineers of fire departments who desire technical information.

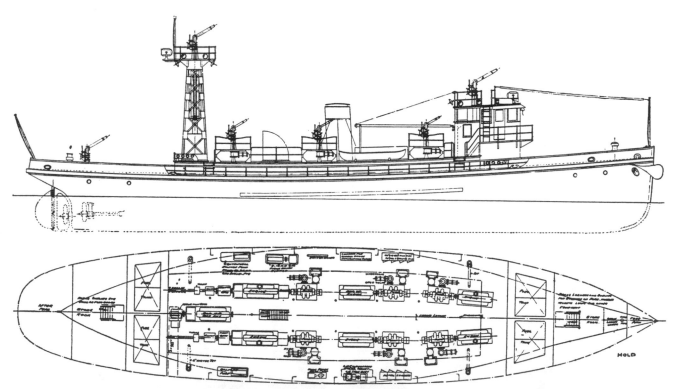

**Outboard Profile and Hold Plan of Gasoline Engine Driven, Centrifugal Pump Equipped Fire Boat**

## PRINCIPAL CHARACTERISTICS:

| | | | |
|---|---|---|---|
| Length overall | 123′ 6″ | Gasoline bunker capacity, gallons | 5500 |
| Length between perpendiculars | 118′ 0″ | Power plant | 7–1200 R.P.M.–300 H.P. engines |
| Beam molded | 26′ 0″ | Pump capacity | 12,000 gallons per minute |
| Beam overguards | 27′ 0″ | Pump pressure | 200 pounds |
| Draft | 7′ 6″ | Monitor capacity | 24,500 gallons |
| Speed, knots | 14 | Salvage pumps | 2–1000 gallon–25 H.P. |

Propulsion is provided by two 500 horsepower General Motors 8-cylinder supercharged diesels. Fuel capacity is 8290 gallons. As rebuilt, the *Alki* is equipped with four underwater maneuvering jet nozzles and a depth finder.

Immediately before the stock market crash and Great Depression starting in 1929, America's fireboat fleet was a hodge-podge of coal, gasoline and diesel-powered craft. Buffalo's *W.S. Grattan*, delivered November 15, 1900, was among the nation's fireboats that continued to make-do with coal-fired boilers and steam-operated piston pumps. Marine-surveyors found the *Grattan's* boilers were dangerously in need of replacement. Their warnings were ignored.

The 9000 gpm *Grattan*, 118-feet-long and with a beam of 24-feet, was among the largest in the United States. It had to be, considering Buffalo's history of grain elevator, flour mill, warehouse and other waterfront disasters. The *Grattan's* crew thought they had seen the worst that fire could do until shortly after midnight, Friday, July 27, 1928, when alarm gongs clanged for a fire at the Atlas Oil Company on the Buffalo River at the foot of Maurice Street, upstream from the fireboat's berth.

As the firefighters pulled on their boots and coats, grabbed their helmets and ran from their 2½-story wooden firehouse to board the *Grattan*, the shimmering glow off to the southeast left no doubt this was going to be a long, difficult night.

At the wheel, Pilot Thomas Hylant guided the single-propeller Swedish iron-hulled *Grattan* as fast as he dared along the winding Buffalo River. Beside him stood the boat's skipper, Lieut. Henry Schickenbantz, who wondered, as did Hylant and the other crewmen, how they were ever going to fight this fire as the loomup worsened.

Below deck, Engineer Thomas J. Lynch coaxed more pressure from the boilers as he got more coal ready for the hours of pumping he knew lay ahead. Gobs of thick black smoke chuffed from the *Grattan's* twin stacks and streamed in its wake. Nearing the fire, the *Grattan's* crew stood ready at the two bow turrets and pilothouse gun as they made ready to engage in battle.

The glare they had seen from their station told them this was going to be at least a three-alarmer. As the *Grattan* came around a sharp bend in the river its crew saw for the first the full magnitude of the fire. The large oil barge, *Cahill*, was a fountain of flames which were threatening to spread into the Atlas refinery and its many storage tanks filled with petroleum products. This was a fire for which the nearly 30-year-old *Grattan* had not been built to fight.

Schickenbantz rang the engine room telegraph as he called for pumping to commence. Engineer Lynch answered by opening valves that fed the two bow turrets and the big cannon on the pilothouse. The pounding pistons of the reciprocating pumps and the roaring rumble of the boilers sent shivers trembling throughout the *Grattan* as its guns bored into the flames.

Even at 9000 gpm, the *Grattan's* firepower was no match for the fire as thick black smoke boiled and rolled thousands of feet into the sky as the glow lit up the city for miles around. Going on 11 hours after the first alarm, the fire continued to stubbornly resist the *Grattan's* streams as well as those of many land-based fire engines, including Engine 12, which had managed to get close enough to take suction from the Buffalo River.

Around mid-morning, Mayor Frank X. Schwab boarded the *Grattan* for an up-front view which was duly-recorded by photographers and newsmen who heard the mayor say how proud he was of Buffalo's bravest. Then he left for City Hall. Shortly thereafter, the *Cahill* broke away from its moorings at 11 o'clock. Fully-ablaze, it drifted across the Buffalo River to the Socony (Mobil) Oil docks at the foot of Babcock Street where it slammed into the huge bulk oil Great Lakes tankship, *McColl*, which instantly ignited to create a worse inferno.

Blazing petroleum products gushing from the *Cahill* and *McColl* turned the Buffalo River into a river of fire. With flames and heat swarming around the *Grattan*, its crew stood their ground. And then a mighty explosion ripped through the *Grattan's* boiler room. Flames swarmed over the pilothouse and deckhouse of the *Grattan*. Leaping overboard, her crew hoped for a miracle that would save their lives. Four of the *Grattan's* crew were severely burned, including Lieutenant Schickenbantz, Pilot Hylant and his brother Fireman John Hylant. The body of the *Grattan's* engineer, Lynch, was recovered after the 17-hour fight to control the fire. Lost in the overwhelming shore-to-shore flames was Engine 12. The *Grattan* was a burned out hulk and, except for her sturdy iron hull, probably would have suffered meltdown.

Everything combustible aboard the *Grattan* was consumed. The fireboat, sitting dead in the water, was little more than a bucket of ashes and other debris. The day after the *Grattan* blew up, a political debate began in City Hall and continued interminably. What to do about the *Grattan*? There was no question Buffalo needed a better fireboat. The city's 42-year-old *George R. Potter*, in reserve and nearing its date with the junkyard, was no longtime answer. But it could serve until the *Grattan's* fate was decided.

A three-way squabble escalated into what bemused taxpayers compared to a Gilbert and Sullivan comedy opera. Mayor Schwab wanted to buy a new triple-duty boat. It would serve as a combination fireboat, police patrol boat and as an icebreaker to keep navigational lanes open and to service the city's Lake Erie water supply Crib intake several miles offshore.

The mayor noted that a police boat should have top priority. Waterfront fires and icebreaking chores were relatively rare. A police boat was needed to snare liquor-runners who freely and every day brought illegal bottled goods from Fort Erie, Ontario, across the Niagara River to Buffalo during those Prohibition years that had started in 1919 and were to continue to 1933 when the 21st Amendment to the United States Constitution repealed Prohibition.

While it must be noted that Buffalonians did not eagerly embrace the idea of cutting off their Canadian liquor connection, it must also be noted that Fire Commissioner George W. Hedden did not either. For the record, however, he argued for a single purpose fireboat. What was more important? Prohibiting whiskey from entering Buffalo and arresting rum-runners or prohibiting fires from destroying the waterfront? Hedden was treading dangerously. Without civil service protection, he served at the mayor's pleasure. For its part, the common council vacillated. Itself failing to provide leadership in settling the matter, the council demanded the mayor show leadership and do something.

The hassle was interrupted more than a year later, November 22, 1929, after the stock market crash and start of the Great Depression. The Duluth, Missabe & Northern Railway offered to sell the city the railway's *William A. McGonagle* which had not only protected their massive wooden trestles for loading iron ore onto Great Lakes vessels, but served Duluth as well.

Anybody who knew anything about fireboats knew the *McGonagle* was a technologically better boat than the *Grattan*. And the railway indicated it would part with the boat for a price that could not be refused. Although the *McGonagle* was only eight years younger than the *Grattan*, it had seen little fire duty and could, therefore, be considered virtually new. The *McGonagle* featured steam-turbine propulsion and pumping. Rated at 12,000 gpm, it could deliver one-third more water than the *Grattan*. The *McGonagle* was, moreover, immediately available.

If anyone had bothered to ask fireboat authorities, they would have recommended that Buffalo grab the railway's offer and get back into full fireboat service immediately, rather than wait the lengthy time it would take to redesign and rebuild the *Grattan* or the longer delay that would result to design and built a new boat from the keel up. Without hesitation, Fire Commissioner Hedden instantly dismissed the *McGonagle* offer. He said his fire department would not accept "second hand junk. We're not in the market for any stuff like that." If Buffalo officials accepted that premise—and they did—then they would have had to concede that the *Grattan* was a dinosaur. Which they did not. The *McGonagle*, sold as a tugboat, served two Ontario ports and, in 1966, was gutted and converted into a pier at Siddal Fisheries, Ont.

The *Grattan's* destiny, meanwhile, continued to be a political football punted from one end of City Hall to the other. Mayor Schwab held out for his triple-duty police boat and Hedden, just as stubbornly, held his steady course for a fireboat. The common council, quietly aware that their thirsty constituents knew how difficult it was becoming to smuggle whiskey past agents along the banks of the Niagara as well as at the Peace Bridge, publicly used economic hard times as an explanation for the lack of definitive action on the *Grattan*.

The only point of mutual agreement seemed to be that the new boat, whatever its purpose, or a rebuilt *Grattan*, would not be gasoline-powered as the state of the art *Scott* in Los Angles and Seattle's *Alki*. Mayor Schwab especially wanted no part of anything fueled by any petroleum product. He well remembered that he had been grandstanding for newsmen on the deck of the *Grattan* shortly before the fireboat blew up; else he might have been killed. It was said that after his self-described "narrow escape" from the *Grattan* he came close to selling his automobile and buying a horse-and-buggy.

Fortunately, no major fires taxed the *Porter* as the political farce dragged on. Realizing that Buffalo's luck could not last forever—and with the election of a new mayor—the city finally advertised for bids to rebuild the *Grattan*. The depression had made thoughts of a new boat totally out of the question. The Buffalo Drydock Company's low bid of $99,500 won them the reconstruction job and the *Grattan* was towed to their yard. On November 18, 1930, the rebuilt *Grattan's* test runs and pumping tests were watched by 3000 spectators close to the boat's berth near the Michigan Avenue Bridge.

*While Buffalo city officials were embroiled in debate over the fate of the city's fireboat, **W.S. Grattan**, which was virtually destroyed during a 1928 oil barge fire and explosion, a railway offered to sell Buffalo the 12,000 gpm technologically-superior **William A. McGonagle**, which was more than one-third more powerful than the **Grattan**. Fire Commissioner George W. Hedden rejected the offer. He said his department would not accept "second hand junk". From the Collection of Steven Lang*

The drydock hospital excised the remains of the old coal-fueled steam boilers, the three steam-operated piston pumps and the steam engine that drove the propeller. Implanted were oil-burning steam boilers and new pumps which did not change the *Grattan's* 9000 gpm capacity. In its exterior configuration, however, there were remarkably new features. While the *Grattan's* deckhouse remained basically unchanged, a new pilothouse was built on the foremost top of it. More striking was the addition of a 22-foot water tower-mounted gun near the stern.

Two 6-inch diameter guns near the bow could discharge streams 600 feet, which especially impressed grain and flour milling owners in Great Lakes' cities. Powerful far-reaching streams are essential when explosions and fires touch off tall grain elevators and flour mills. How could the grain milling establishment have foreseen the day—30 years later—when Buffalo's fireboat would lay claim to a unique experience in the history of the American fireboat fleet? The Buffalo boat, without suitable open sea navigational equipment and in the dead of night, crossed international waters to battle and control a mammoth flour mill fire in Canada and safely returned to Buffalo where crewmen and other firefighters recounted a story that became legendary in the history of the Buffalo Fire Department.

With the recommissioning of the *Grattan*, November 20, 1930, more than two years after she had blown up, the 44-year-old coal-burning, piston-pumping *Potter* went into a well-deserved retirement. But the metamorphosis of the *Grattan* did not end with her resurrection. In 1953, the *Grattan* voyaged to the Sturgeon Bay Shipbuilding & Drydock Company, Sturgeon Bay, WI., for a complete facelift. Tests had shown the *Grattan's* boilers could only raise pressure to provide 40 per cent of the rated capacity.

Rebuilt after the 1928 explosion and fire which left it a bucket of ashes, Buffalo's **W.S. Grattan** returned in 1930 to its berth at Engine 20 on Ohio Street near the South Michigan Avenue Bridge. The rebuilt fireboat was notable for its conversion to oil-burning steam boilers. A new pilothouse was built over the deckhouse and a 22-foot water tower gun was mounted near the stern. Buffalo Fire Historical Society

After more than half-a-century in fresh water, the Swedish iron hull remained shipshape. Four Model D397 Caterpillar Tractor Company of Peoria, Ill, V-type 12 cylinder engines, each rated at 400 horsepower, were installed. To provide propulsion to the two new cast steel blades, 5-feet in diameter, at a normal speed of 300 rpm. The two Caterpillars forward of the propulsion powerplants are coupled to four 3750 gpm pumps built by Dean Hill Pump Company of Indianapolis, thus beefing up the *Grattan's* heft to 15,000 gpm. Two Caterpillar D311 diesel electric generators furnish auxiliary power. The boat's twin stacks were replaced by stubby diesel exhaust vents and the stationary tower gave way to one that could be hydraulically raised 15 feet above deck.

Returning to Buffalo on December 2, 1953, the old *Grattan* got a new name: *Firefighter*. Unlike the fireboat, that name was short-lived. With the 1954 death of Edward M. Cotter, president of Buffalo Fire Fighters Local 282, AFL-CIO, the boat was renamed in his memory. Surviving a shattering explosion and fire, political shenanigans, countless fires, fierce winters and name changes, Buffalo's fireboat was, in 1989, closing fast upon its 100th birthday. The *Edward M. Cotter* continues to be the oldest active fireboat in the history of the United States and is likely to remain so during the foreseeable future as the antique boat with a modern-day look stands ready to meet whatever challenges come its way.

An unparalleled voyage, unique in American fireboat history was the **Edward M. Cotter's** dead-of-the-night crossing of domestic and international waters to fight a disastrous fire. Port Colborne, Ont., officials, without fireboat protection, had appealed to Buffalo Fire Commissioner Robert Howard for help when fire erupted in the eight-story Maple Leaf Milling Company grain and flour milling complex. The **Cotter's** crew, augmented by land-based Engine 8 firemen, left Buffalo at 8:30 Friday night, October 7, 1960, and headed due west by north for the 30 statute mile trip across Lake Erie. Lacking radar and suitable aids for navigating the often treacherous lake, the 60-year-old **Cotter** was escorted by a U.S. Coast Guard cutter. Arriving in Fort Colborne more than two hours later, the **Cotter's** guns were brought to bear at 10:46 p.m. For more than four hours the fireboat's 15,000 gpm at 150 psi capacity fed nozzles battering the flames. Bringing them under control at 4:10 a.m., Saturday morning, the **Cotter** returned to Buffalo. Buffalo Fire Historical Society

*Modernized with diesel-electric power and with its pumping capacity increased to 15,000 gpm, Buffalo's fireboat **Grattan** was renamed **Edward M. Cotter** in 1954. Built in 1900 and retaining its original Swedish iron hull, the **Cotter** is the oldest active fireboat in the history of the United States. From the Collection of Bill Dahlquist*

*All seven monitors of Tacoma's 10,000 gpm **Fire Boat No. 1** battle flames sweeping the old London Dock in 1936. Washington State Historical Society*

During the years of prosperity leading up to the stock market crash and the Great Depression starting in 1929, larger and more fireboats created what can only be described as a boomlet in this relatively small segment of the boatbuilding industry. From 1919 to the start of the depression, fire departments bought 19 large fireboats. During the economic hard times that ended around 1941, only seven fireboats were built. The era of bigger boats meant bigger investments—a hard sell in the best of economic times. In a depressed economy with firemen being laid off, or paid in script, new fireboats were red-inked out of city budgets and considered extravagant luxuries.

In this gloomy economic atmosphere New York opted to purchase what was described—not always complimentarily—as the Rolls-Royce of fireboats. Its $982,784 cost far exceeded any fireboat built up to that time. Why did New York need another fireboat when it had a fleet of nine? It had 10 boats until September 4, 1934, when the 52-year-old *Zophar Mills* was taken out of service. Another possible justification for a new boat was that the *William L. Strong,* at 40, only had a 6500 gpm capacity; half that of most other newer American fireboats.

On the other hand, the pride of the FDNY Marine Division was its newest boat, *John J. Harvey,* only several years old. This most powerful gasoline-operated fireboat ever built had a 16,000 gpm at 150 psi capacity. Hypothetically, moreover, New York's nine fireboats could, at that time, simultaneously deliver 77,500 gpm. New York's 600 miles of waterfront, around 400 piers and some 11,000 ocean-going vessels calling upon America's biggest seaport, created the probability of major fires. Even with that potential for disaster, most authorities considered New York far better fireboat-protected than any other city in the world.

*The foredeck monitor of Tacoma's **Fire Boat No. 1** packed a 10,000 gpm punch. Dr. Allen W. Ratcliffe of the Tacoma Landmarks Preservation Committee said the Big Bertha gun "could peel off the roof of a burning building and get better access to a fire." Washington State Historical Society, Tacoma*

*Destined to become one of Chicago's most famous fireboats, the **Fred A. Busse,** named after a mayor, went in service May 1, 1937. Built at Defoe Shipbuilding Company, Bay City, Mi., the **Busse** was rated by the National Board of Fire Underwriters at 10,000 gpm at 150 psi. The steel boat had four turret guns: two near the bow, one toward the stern and a tower-mounted gun which could be elevated 27 feet. The **Busse** had four Dean-Hill three-stage centrifugal pumps and was operated by six diesels. Weighing 157 gross tons, the boat was 90¹/2-feet-long, with a breadth of 22.4-feet and a 7.1-foot draft. The **Busse** had quarters for a crew of eight. Its low profile was typical of fireboats built for cities on the Great Lakes because they could pass under pier bridges without waiting for them to be raised. During its 44 years of service, the **Busse** fought many fires, including this 5-11 alarmer in the summer of 1966 at a box factory, formerly the Rheingold Brewery, at 3501 Elston Avenue. The **Busse's** final trip, October 7, 1941, had its 20-year pilot, Willie Fridell, at the controls as the boat went into retirement at the 135th Street drydock on the Calumet River. Steve Little from the Collection of Ken Little and Bob Freeman*

*Trial run in October, 1929, of the first single purpose fireboat of Tacoma, Wa., **Fire Boat No 1,** built at the Coast Line Shipbuilding Company, Tacoma. The 10,000 gpm at 150 psi all-steel boat cost $148,000. Displacing 88 tons, Tacoma's boat was 96¹/2-feet-long, 21¹/2-feet in breadth and had a 6-foot draft. As built, the fireboat had four Sterling-Viking, 425 horsepower, gasoline-operated engines, which were variably used for propulsion and pumping. The engines drove four De Laval three-stage centrifugal pumps, each rated at 2500 gpm. The triple-propeller craft was said to be capable of speeds exceeding 14 miles per hour. Washington Historical Society, Tacoma*

Dripping with icicles, Baltimore's 12,000 gpm **Torrent** returns to its berth after battling a six-alarm fire, February 21, 1936, aboard the berthed Norwegian freighter, **Gisla**, loaded with 5000 tons of nitrate. The 16-hour firefight in sub-freezing weather killed one fireman. Built in 1921 as a coal-fueled, steam piston-

pumping boat, the **Torrent** was dieselized in 1936. The steel **Torrent** had two Worthington centrifugal pumps, two double-acting Ahrens-Fox pumps and five monitors. It was 121-feet-long, had a 29½-foot beam and a 12½-foot draft. *Charles Cornell*

Boston fireboats quickly attacked, March 10, 1937, when five alarms were sounded for a fire aboard the freighter **Laila**, at Pier 45. The 42-year-old **Angus J. McDonald** and other fireboats lobbed streams into the fully-involved **Laila**. The

**McDonald** remained in service for another decade. *From the Collection of Bill Noonan*

Perhaps we get to the nub of the mystery by focusing upon the two movers and shakers who were most responsible for the decision to build a fireboat without peer. Marine Architect William Francis Gibbs, a partner in New York's Gibbs & Cox, was as avidly interested in firefighting as he was in marine design. To the shipping world he would best be remembered as the designer of the ocean liners *America* and *United States*, and World War II's cargo-carrying Liberty Ships. To New York firemen he was one of countless fire buffs, those sidewalk superintendents of firefighting.

Gibbs, super salesman and super buff, knew Mayor Fiorello H. Laguardia; himself as devoted a buff as he was skilled at getting what he wanted from City Hall politicians. Whether the super fireboat idea originated with Gibbs, which is likely, or with Laguardia, which is possible, may never be known. How Laguardia justified the expenditure is not known, either, except that it would bring employment to the city's depressed shipbuilders and suppliers. What is known is that Laguardia coaxed the city fathers into appropriating money in 1937 for what would become the most famous fireboat in American history.

Gibbs designed the all-steel fireboat the way he did ocean liners. Big. His plans called for a diesel-electric boat—New York's first—with a firepower of 20,000 gpm; more than

*Chicago's **Fred A. Busse** (left) and **Joseph Medill** averaged 11,000 gpm while battling the May 11, 1939, explosions and fires in five elevators of the Norris Grain Company and Rosenbaum Brothers, Inc., along the Calumet River on the city's south side. Forty engines and 20 other fire companies battled the blaze which started with an explosion at 8:52 a.m., and led to one of the worst grain elevator*

*disasters in Chicago's history. A falling wall damaged the **Medill**'s deckhouse and snapped off the starboard turret gun, but the boat quickly rejoined the battle and pumped for 50 hours. The **Busse** pumped for 53 hours and 35 minutes, said to be an all-time record for Chicago fireboats. From the Collection of Paul Ditzel*

double that of some of the Marine Division's boats. Specifications included a length of 134-feet, a 32-foot beam and a 9-foot draft for a boat displacing nearly 600 tons. The reinforced steel bow would enable the boat to quickly plow through ice. A heating system not only kept the machinery warm for quick use, but prevented freeze-up of the manifolds. Standing low in the water, the fireboat would easily pass under any New York bridge and maneuver in deep water as well as the city's many shallow inlets.

In designing the boat, Gibbs drew upon diesel-electric enthusiasm that was revolutionizing American railroading as the age of speedy diesel-powered transcontinental streamliner trains tolled the end of steam locomotives. Gibbs adapted this technology, also newly-used in Navy submarines, into his design. There is no doubt that Gibbs' widely-publicized diesel-electric fireboat hastened the end of steam turbine and gasoline power in the American fireboat service.

New York's boat, officially *Hull 856*, was the last built at the depression-plagued United Shipyards, Inc., Mariners Harbor, Staten Island, before the company was acquired by Bethlehem Shipbuilding Corp. As the launch date approached, Eleanor Grace Flanagan, 18, daughter of Lieut. Joseph Flanagan of the fireboat *George B. McClellan*, was chosen to christen the new boat. She won the honor because her 95.17 scholastic record was the highest of any fireman's daughter graduated that year from a New York City high school.

On August 26, 1938, Miss Flanagan stood ready with a bottle of champagne at the bow of *Hull 856* along with Fire Commissioner John J. McElligott, Gibbs, Mayor Laguardia and some 1000 invited guests, including the FDNY band. Of all the VIPs, Miss Flanagan's presence was remarkable for her white satin ensemble decorated with fire engine red pom-poms, cloak and simulated fire helmet.

The brunette smashed the champagne across the bow and christened the boat *Fire Fighter* as it slid down its cradle and splashed into the waters of the Kill Van Kull. The guns of the *William J. Gaynor* and *John J. Harvey* gushed a water salute as nearby vessels hailed the *Fire Fighter* with shrieking whistles.

Enter another mystery. If FDNY tradition prevailed, the boat would have been named after a mayor. Considering that LaGuardia was New York's most famous mayor and devoted to the fire department as well, naming the boat *Fiorello H. LaGuardia* would have seemed appropriate. Perhaps the answer lies in criticism of LaGuardia for his indulgence in an expensive purchase—and a fireboat at that—when unemployed New Yorkers were selling apples on street corners.

Laguardia's explanation was that the *Fire Fighter's* name was selected to honor all New York firemen. Perhaps. But there can be no question that Laguardia, modesty not being his strongest suit, lusted for a fireboat with his name on her bow. The *Fire Fighter* went in service at 9 a.m., November 16, 1938, at the Battery at the southerly tip of Manhattan

*New York's **Fire Fighter**, the most formidable fireboat ever built, was to become almost as famous as the Statue of Liberty shown off her port bow. When new, as shown, the 20,000 gpm at 150 psi **Fire Fighter** mounted nine guns of between* 2000-6500 gpm capacity. The large deck pipe at the bow could deliver 27 tons of water a minute. The water tower, a frequent source of mechanical problems, was removed in 1962. From the collection of Peter E. Balducci

Marine Wiper Arty McCrossen stands aft of the two hulking submarine-type Winton/Cleveland (General Motors) diesel engines which operate New York's **Fire Fighter**. The engines, mounted amidship, are each rated at 1500 horsepower at 750 rpm. Behind McCrossen are the six power generators which supply direct current for operating the engines, pumps and other needs of the boat. Robert M. Brewis

Two of the **Fire Fighter's** four Westinghouse 60 horsepower at 1500 rpm motors which drive the New York fireboat's four forward-mounted 5000 gpm De Laval two-stage centrifugal pumps which deliver 20,000 gpm at 150 psi. Directly above FDNY Marine Engineer Danny Reddan are butterfly valves; part of the fire main control system. The **Fire Fighter's** water intake is from a T-shaped, 27-inch diameter tunnel which can draw from the hull on the port and starboard sides, or the bottom of the boat. The pumps discharge into a 14-inch diameter loop main connected to eight deck pipes and 21 gated manifold outlets. Robert M. Brewis

*The **Fire Fighter**'s engine room operating console. Robert M. Brewis*

Island. The fireboat was a major attraction during the 1939 New York World's Fair whose theme, The World of Tomorrow, celebrated technology as epitomized by the *Fire Fighter*.

As massive as the behemoth is in profile, the *Fire Fighter*, referred to by her crew as the *Fighter*, is even more impressive below deck where power generators, propulsion, pumping and related systems are compactly nestled. Powering the *Fire Fighter* are two submarine-type Winton/Cleveland (General Motors) 16 cylinder, V-type diesel engines, each rated at 1500 horsepower at 750 rpm. The fuel supply, sufficient to operate the boat for a week, is contained in fore and aft bunkers with a total capacity of 34 tons. The oil is centrifuged and cleaned before reaching the engines.

Immediately aft of each huge, side-by-side, center-mounted diesel engine are three generators which supply direct current. They have varying capabilities. One generator accommodates one propulsion motor or two fire pumps. The other generator not only supplies one or both propulsion plants, but can be used to provide air compressor power to jackhammers for breaching concrete pier decks, as well as chain saws and other tools. The third generator serves as an exciter, or dynamo, for the two other generators and provides on-board lighting as well as auxiliary power. The *Fire Fighter* is outfitted with a 1,500,000 candlepower searchlight and five 1000 watt floodlights; all pilothouse-controlled.

Evidence of Gibbs' pioneering design is the *Fire Fighter's* direct propulsion control from the pilothouse, instead of the traditional telegraph systems for relaying orders to the engine room. A shaft directly connects each propulsion motor to a 6-foot bronze, three-bladed propeller. With two rudders for better control, steering is via an electro-hydraulic system. Top speed is about 16 miles per hour.

Four 5000 gpm De Laval two-stage centrifugal pumps mounted forward of the gigantic diesel engines, provide the *Fire Fighter's* 20,000 gpm at 150 psi capacity. A Westinghouse 600 horsepower at 1500 rpm motor powers each 5000 gpm pump. When firefighting calls for more pressure, the De Lavals operating in series, deliver 10,000 gpm at 300 psi.

The pumps draw water from a 27-inch diameter T-shaped tunnel in the hull. Suction valves enable the pumps to draw from port, starboard or the bottom of the boat. The pumps discharge into a 14-inch diameter loop main connected to eight deck pipes and 21 gated manifold outlets for 3½ inch hose.

The *Fire Fighter's* exterior profile is notable for its larger-than-normal pilothouse and the relatively long deckhouse abaft of it. Probably borrowing an idea that originated with Los Angeles' *Ralph J. Scott* fireboat, Gibbs included a perforated pipeline to provide water spray protection against radiated heat and flammable liquid fires floating on the water.

Directly below the pilothouse is what has become known as the Gold Room for its storage of various sizes and types of brass nozzles and related fittings. While the *Fire Fighter* was never intended to provide sleeping accommodations for its entire crew, Gibbs designed a modest-sized crew quarters. The pilot sleeps on a couch in the pilothouse. The two marine engineers and a marine wiper also sleep aboard the *Fire Fighter*. This is mainly a safety factor in the event that unusually strong tides or storms cause the *Fire Fighter* to break loose from its berth and, adrift, become a hazard.

The *Fire Fighter's* crew assignment has varied over the years, largely because of staffing costs. In 1988, the *Fire Fighter's* on-duty crew consisted of a pilot, two marine engineers, a marine wiper, four firefighters and an officer-in-

The **Fire Fighter**, flagship of the FDNY Marine Division fleet, demonstrates its 20,000 gpm at 150 psi capability with this water display usually reserved for welcoming new ships arriving in New York on their maiden voyages. By syphoning environmentally approved food dye granular powder with a pickup tube into each nozzle the resulting venturi effect enables the **Fire Fighter** to create patriotic fountains of red, white and blue water. The diesel-electric powered boat is 134-feet-long with a beam of 32-feet, a draft of 9-feet and displaces nearly 600 tons. From the Collection of Paul Ditzel

charge. That crew can be augmented by firefighters from land companies. When the *Fire Fighter* answers calls for help outside New York City, operating procedure calls for a battalion chief, usually from the Marine Division, to accompany the boat.

Gibbs took advantage of the length and height of the pilothouse to design a monitor platform extending from the pilothouse to the diesel exhaust stack where, like a horseshoe, the platform fits around it. Mounted on this platform are five deck pipes ranging from 2000-3000 gpm capacity. Two more guns, abaft the stack, are mounted on the deckhouse. But it is the *Fire Fighter's* bow deck pipe that packs the most wallop. With a 5-inch diameter nozzle, the gun can put out 6500 gpm: 27 tons of water a minute.

A ninth deck pipe of 3000 gpm capacity was mounted on a steel tower designed by Gibbs. Normally cradled horizontally and pointing to the stern, the tower was hydraulically raised 55-feet above water level. First of its kind, the tower was removed in 1962, mostly due to frequent breakdowns and repair costs. Further enhancing the *Fire Fighter's* capabilities are the port and starboard gated manifold outlets for coupling to hose which can supply land-based fire companies. Hose is weather and fire-protected on three brass-plate-covered reels, one fore and two near the stern.

For fighting fires under wooden wharves and piers, hose-supplied streams can be operated from hull portholes. The *Fire Fighter* is outfitted with two Boston Whaler boats and an inflatable rubber boat. These are more likely to be used for

*The **Fire Fighter**'s longest run occurred in 1980 when the New York fireboat traveled under her own power to Robert E. Derecktor's shipyard in Coddington, R.I., for replacement of aging fire mains. After sandblasting and other standard drydocking services, the **Fire Fighter** returned to duty. James Murray*

*Removal of the **Fire Fighter**'s poorly-functioning water tower in 1962 and the addition of a radar mast are about the only exterior changes noticeable during the New York fireboat's first 50 years of service. Since May 12 1973, the **Fire Fighter** has been berthed near Staten Island's St. George Ferry Terminal which burned on*

*June 25, 1946. The nine-alarm fire required six fireboats, including the **Fire Fighter**, and 50 fire companies. Three persons were killed and 250 firefighters and others were injured. From the Collection of Paul Ditzel*

The *Fire Fighter*'s crew prepares for the New York fireboat's golden anniversary celebration in 1988. A popular fireboat among modelers, the *Fire Fighter* has had five color schemes. When new, the boat had a black hull and water tower, white deckhouse and pilothouse. Its stack was buff-colored with a black band around the top. Fittings and nozzles, manifolds, reel covers and bitts were highly-polished brass. By 1988, the *Fire Fighter*'s hull below the waterline was black. Fire engine red extends from there up to and including the fenders. The outer bulwarks are white and the inboard are red. The pilothouse and deckhouse are white. The deck and monitor platform are fire engine red as are the guns, stack and radar mast. Black trims the top edge of the stack as well as the top edge of the bulwark rail, hose reel covers, bitts and railwork around the monitor platform. Robert M. Brewis

This colorful crew patch celebrated the Fire Fighter's golden anniversary. The red, white and black boat was set against an orange background, with aqua in the lower third of the Maltese cross. F.D.N.Y. was in royal blue and 50th was in gold. All other wording was in black. Bill Guido

*Logbooks of the **Fire Fighter** are a history of New York area waterfront firefighting and rescues. With its tower and deck guns operating and a port side gun directed down to the water to help hold the boat on station due to back pressure of the other nozzles, the **Fire Fighter** is joined in battle by the 7000 gpm **Abram S. Hewitt**. Considering that the **Hewitt** was retired in 1958, this fire occurred between 1938 and 1958. From the Collection of Bill Noonan*

underwharf and pier firefighting, search, rescue and incidents in small boat marinas where the *Fire Fighter* cannot be taken.

Gibbs designed petrochemical firefighting capability into the boat by outfitting it with four portable foam generators with a 500 gpm capacity at 100 psi. The generators were later removed and newer petrochemical firefighting equipment was installed. Other features built into the boat included a loudspeaker system, fire and Coast Guard radio communications, radar, a fathometer and engineroom ventilation as a protection against smoke intrusion. A carbon dioxide fire extinguishing system helps to safeguard the engineroom.

During her first half century of service, the *Fire Fighter's* journals read like a compendium of major waterfront fires in and around New York. She fought her first major fire, January 23, 1939. At 2:30 a.m., the *Fire Fighter* answered a call to the foot of 57th Street in Brooklyn where, for 13 hours, the *Fire Fighter* supplied 30 hoselines used to battle a four-alarm fire aboard the *Silver Ash* loaded with rubber.

On December 3, 1956, the *Fire Fighter* was the first fireboat to arrive at the burning Luckenbach Steamship Pier at the foot of South Brooklyn's 35th Street. As she moved in to attack, an explosion on the 1700-foot-long dock and buildings badly damaged the *Fire Fighter*. Many of her crew were blown overboard and injured. Ten persons were killed and

250 hurt as fire swept the Luckenbach pier.

Four years later, the *Fire Fighter* was one of four fireboats called to the December 19, 1960, deep-seated fire aboard the aircraft carrier, *Constellation*, nearing completion in the New York Naval Shipyard, Brooklyn. The boats supplied water to firefighters battling one of the most treacherous and punishing fires in FDNY history. During the 17-hour battle, 50 shipyard workers were fatally injured and 336 injured.

On May 12, 1973, the *Fire Fighter*, officially Marine Company 9, found a home at a slip near the St. George Ferry Terminal, Staten Island, where she would remain. It was from there, six years later, that the fireboat would speed to a catastrophic ship collision and fire near the Verrazano-Narrows Bridge.

The courageous feats and skills of the fireboat's crew that horrendous night earned the *Fire Fighter* The Gallant Ship Award, the United States Government's highest commendation as well as the American Merchant Marine Seamanship Trophy for distinguished seamanship and heroism.

But trophies were not uppermost in the minds of the *Fire Fighter's* crew, late in 1941, as they prepared for possible American entry into World War II. The flagship of the FDNY Marine Division fleet stood ready at the entrance to New York Harbor for a host of new problems that could arise in the port or from enemy air and sea attack.

The Cunard Line fire at Pier 54, North River, was the first waterfront disaster fought by New York's **John J. Harvey,** the largest and most powerful gasoline-operated boat built for the American fire service. The five-alarm fire, starting at 6:53 a.m., May 5, 1932, at the steamship line's West 14th Street and the North River terminal, was one of the worst pier fires in FDNY history. The **Harvey,** built in 1931 at Todd Shipyards Corporation, Brooklyn, was designed by Henry Gislow, Inc. The 16,000 gpm steel boat cost nearly $600,000 and is 130-feet-long, with a beam of 28-feet, a draft of 9-feet and weighs 268 gross tons. As built, the **Harvey's** five main engines were connected to five 340 kilowatt generators which powered two propelling motors as well as four LeCourtney pumps. The **Harvey** has eight monitors, including unconventional dual monitors on a tower near the

stern. Commissioned December 31, 1931, the boat was named after John J. Harvey, pilot of New York's fireboat **Thomas Willett.** He was killed, February 11, 1930, while the Willett was helping to fight a five-alarm fire aboard the **SS Muenchen.** The 526-foot-long freighter was loaded with explosive chemicals, shellac and newsprint while berthed at Pier 42 on the North River at the foot of Morton Street. A violent explosion aboard the **Muenchen** destroyed the **Willett's** pilothouse. Harvey was killed and several fireboat crewmen were blown overboard. The **Harvey,** dieselized in 1957, is still in service after 57 years, a record unsurpassed in FDNY Marine Division history. Note the close exterior profile resemblance to the **Fire Fighter,** which was commissioned seven years later. From the Collection of Bill Noonan

The twin-stacked **Abram S. Hewitt**, along with the **John J. Harvey**, were among most of the FDNY's Marine Division fleet called to help battle the Cunard Line fire at Pier 54 and the North River on May 5, 1932. Burning creosoted wood pilings under the concrete dock supporting the three-story Cunard transoceanic terminal created toxic and skin-irritating smoke conditions. The five-alarm fire required recall of off-duty firemen. Of the 850 firefighters who battled the fire, 300 were injured, mostly from smoke inhalation. Especially noteworthy in this photo are the two streams shooting from portholes near the starboard bow. As fireboat technology progressed, porthole-operated streams gave way to hull-mounted underwharf and pier firefighting nozzles and remotely-controlled from the pilothouse. From the Collection of Bill Noonan

# CHAPTER SIX
## "WE'RE AT WAR!"

The crew of the U.S. Navy Yard Tug, *YT-146*, the *Hoga*, was off-duty that tranquil Sunday morning, December 7, 1941, in Pearl Harbor, headquarters of the Pacific Fleet. Quartermaster Bob Brown, 28, second in command of the combination tug and fireboat, was sleeping on a cot in the pilothouse, while nine other crewmen were in the aft deckhouse.

The *Hoga*, a Sioux Indian word meaning "fish", was one of around 100 U.S. Navy vessels, including seven battleships, nine cruisers and more than 30 destroyers—half the United States' Pacific Fleet—nestled inside the naval base. The *Hoga* was berthed with several other tugs at Yard Craft Dock, a few hundred yards across the channel from Battle-

ship Row where 11 vessels, including the battleships *USS Arizona* and *USS Nevada*, were moored at quays just off Ford Island. Many of the vessels along the row were two abreast.

Brown and thousands of other Navy, Army, Marine and civilian personnel at Pearl Harbor and surrounding military airfields were blithely unaware of the death and destruction that was zeroing in on the island of Oahu. Some 220 miles north of Pearl Harbor steamed a Japanese task force of 33 warships, including six aircraft carriers. The first wave of 185 attackers—fighter planes, high-level bombers, dive bombers and torpedo planes—had lifted off from the carriers at 6 a.m. The second wave, bringing the total assault force to 354 aircraft, followed shortly thereafter.

*The battleship **USS Arizona**, battered by bombs and a torpedo which blew up its powder magazine during the sneak attack on Pearl Harbor, sank in less than nine minutes. More than 1000 of her officers and crew were killed. The U.S. Navy yard tug and fireboat, **Hoga**, battled fires aboard the **Arizona** for 48 hours. **USS Arizona** Memorial, National Park Service*

The keenly planned multiple assault included air installations on Ford Island, Hickam and Wheeler Fields, Kaneohe Naval Air Station and the Marine Air Base at Ewa. American planes, most of them standing wingtip-to-wingtip to better guard against sabotage, would be easy prey. Japan's failsafe tactics called for their destruction or damage before they could get off the ground.

Number One on the Japanese hit list was Battleship Row and the 6½-square mile Pacific Fleet Headquarters Base inside Pearl Harbor. Japanese intelligence—faulty, as it turned out—said the Pacific Fleet's three aircraft carriers were also sitting duck targets. If Pearl was struck a devastating blow it would surely destroy the United States' naval and air effectiveness in the Pacific before war was officially declared.

At 7:57 a.m., Brown was jolted awake by the sounds of low-flying, fast approaching aircraft. He thought this was just another routine war games exercise. Looking from the *Hoga's* pilothouse, Brown saw the flaming red Rising Sun emblems under the wingtips of Japanese torpedo bombers passing over the *Hoga*. Battleship Row was dead ahead of them. Swooping low—no more than slightly over 100 feet above water level—Brown stood in disbelief as torpedoes dropped from the bellies of the attackers. Swishing across the channel, the torpedoes bored into Battleship Row as explosion upon explosion erupted into billowing clouds of thick, black smoke and flames. Overhead, bombs rained upon Battleship Row as other planes sighted in on their targets.

The *Arizona*, a short distance up the channel from the *Hoga*, swallowed an unknown number of torpedoes and bombs which mercilessly pelted the battlewagon and exploded. Japan's ace bombardier, Petty Officer Noburo Kanai, in a high-altitude bomber released a 1760-pound, 16-inch-diameter armor-piercing bomb which hit near the *Arizona's* No. 2 turret gun, burrowed to the forward powder magazine and exploded. The tremendous blast lifted the *Arizona* and sent a thick, towering cloud of dark reddish-brown smoke jetting thousands of feet into the sky. The *Arizona* sank in 38 feet of water in less than nine minutes. Only 289 of her officers and crew survived.

Other bombs and torpedoes found their targets along Battleship Row. The explosion-shattered *USS California* began listing and settled to the muddy bottom of Pearl Harbor. The *USS West Virginia* was destroyed by torpedoes and two bombs. With flames swarming over her superstructure, the *West Virginia* followed the *California* and the *Arizona* to the bottom. The torpedoed *USS Utah*, a state of the art, remote-controlled target ship and gunnery training vessel, capsized with the loss of around 63 officers and crew. It would never be raised.

The *Nevada*, tucked at the northern end of Battleship Row was both lucky and unlucky. Lucky, because massive clouds of black smoke boiling from the *Arizona*, immediately ahead of her bow, obscured the *Nevada* from the attackers.

Unlucky, because flames floating on top of the water as burning oil from the hemorrhaging *Arizona's* 9630 tons of oil spilled from her slashed belly, endangered the *Nevada*. Too, the hail of bombs was not selective. The *Nevada* was repeatedly hit, burst into flames in her forepart, and suffered many casualties. The *Nevada* may have been hobbled by the bombs, but it was still able to fight back. Her gunners would be credited with downing two planes.

Chief Boatswain's Mate Joe B. McManus, 31, the *Hoga's* skipper, was aft in his captain's quarters and shaving when the attack started. He ran out on deck and looked across Battleship Row where the *USS Oklahoma*, hit by five torpedoes, was starting to roll over on her starboard side with the loss of 315 officers and crew,. Marion Minehart, the *Hoga's* chief machinist's mate, jumped onto Yard Craft Dock to try to see and comprehend what he refused to believe. "Good God!" he said. "We're at war!"

The Yard Craft dockmaster ordered the *Hoga* to get underway and, according to her log, "assist all ships and pickup survivors." The hours ahead not only challenged the crew of the *Hoga* but put them in harm's way as undreamed of demands were made. In its favor was the fact that the *Hoga* was just under a year old. Launched December 31, 1940, at Consolidated Shipbuilding Corporation, Morris Heights, N.J., McManus and Brown were among those who brought the *Hoga* to Pearl as part of a convoy which encountered turbulently high seas that nearly capsized the boat. The *Hoga* went in service at Pearl's 14th Naval District Headquarters, May 22, 1941, less than seven months before the surprise attack. Until then, the *Hoga's* mundane duties included assisting ships in and out of Pearl and towing and helping to dock vessels around the 12 miles of the base's facilities.

The welded steel and diesel-powered *Hoga* displaced 350 tons and is registered at 99.7-feet-long, with a 25½-foot beam and a 10½-foot draft. The *Hoga*, like many Navy tugs, was outfitted for firefighting, but not for the monstrous fires burning that day. Twin 250 horsepower electrically-operated pumps gave the *Hoga* 4000 gpm capacity at 150 psi. Its firefighting equipment included three monitors, one of them atop the pilothouse, and four on-deck manifolds on the starboard quarter of the deckhouse for coupling to firehose carried by the *Hoga*.

As the *Hoga* backed into the channel that morning, Brown said, "It's a madhouse out there!!" His memory is as vivid today—nearly 50 years later—as it was that day the firefighting tugboat went to war. Battleship Row was a maelstrom of explosions, rocketing clouds of impenetrable black smoke and balls of fire erupting from the warships. Tens of thousands of gallons of oil bleeding from the stricken warships spread gigantic smears into the channel where they flashed into flames surrounding ships along Battleship Row.

Carnage was everywhere as explosions flung bodies hundreds of yards onto land, into the water and onto the decks of other ships. As more bombs splattered Battleship Row, other missiles, missing their targets, exploded in the channel and scooped waterspouts spewing hundreds of feet into the air.

Into this unbelievable hell churned the *Hoga*. McManus saw they must immediately help the repair ship, *USS Vestal*, which had been moored next to the *Arizona*. The *Hoga's* crew got a towline on the *Vestal*, damaged by two bombs and her hull shattered by flying debris from the *Arizona*. McManus could not help a moment's reflection upon his recent assignment aboard the *Vestal* and his promotion which resulted in his transfer to the *Hoga*. The repair ship was hauled to the nearby Aiea Bay mudflats where the *Vestal* was beached, but later sank.

Many bombs hurtling upon Pearl exploded as close as 150 feet from the *Hoga*. It seemed ironic that the warships in Pearl had the firepower to strike back and many of them were firing upon the attackers. But the *Hoga*'s armament shot only water and none of the crew carried sidearms. They found some comfort in the realization that a tugboat, inconsequential compared to the enemy's primary targets, was unlikely to be singled out for attack. While the *Hoga* was not a priority target for the two waves of attackers with their bombs and strafings, the fact of the matter was that bombs, machinegun fire and falling shell fragments were often indiscriminate and could easily hit the *Hoga*, sink her and slaughter her crew.

The *Hoga* continued its work under the ominous dronings of still more attack planes as the warships mounted counterattacks with machine guns and 5-inch shells. The sky over the *Hoga* was pimpled with puffs of black smoke marking exploding antiaircraft shells. The Japanese lost 29 planes during the two-hour attack; a relatively small price to pay for the devastation including eight battleships damaged or sunk; three light cruisers and three destroyers racked by bombs or torpedoes, plus 188 Army, Marine and Navy aircraft destroyed.

At full-ahead 14 knots, the *Hoga* raced to the aid of the *Oglala* and helped the damaged minelayer into 1010 Dock. But it was too late for the *Oglala*. The old, wooden-hulled minelayer capsized. Heading out again toward the burning and smoke-shrouded Battleship Row, where a steady chatter of shells streaked skyward, the *Hoga*'s crew spotted two sailors bobbing in the channel. They had either jumped or were blown off ships. One of them suffered a gaping wound in his leg. The other appeared unhurt. The *Hoga* put both survivors ashore at Berth 3 where medical aid awaited them.

Despite its severe damage, fires and casualties, the *Nevada* got underway 40 minutes after the first attack. Under normal conditions it would have taken around two hours to get up steam and be assisted by tugs as it headed to sea. To get there, the crippled *Nevada* would have to run a gauntlet of bombs, torpedoes and strafings as it lumbered, its bow settling lower as it took on water, from the northernmost end of Battleship Row, down the channel and out the narrow opening leading to the ocean. Passing about 50 feet from the blazing *Arizona*, the *Nevada* picked up two survivors, while its gunners shielded their shells with their bodies out of fear the ammunition would explode from radiant heat.

Japanese attackers immediately saw that one of their primary targets was escaping. Twenty-one planes attacked the *Nevada* with a vengeance that seemed worse than their assaults upon the other warships along Battleship Row. A torpedo exploded amidship on the *Nevada*'s port side and blew a hole "large enough to drive two big trucks through," recalled Quartermaster Second Class Roy G. Johnston, 21, who, as assistant navigator, was surrounded by thick steel in the battleships conning tower. As it made its desperate run down the channel the *Nevada*'s crew fought back with their 5-inchers and machineguns while repeated strafings exacted a heavy toll among her crew.

The damage assessment aboard the *Nevada* was ominous. In addition to the torpedo, the battleship had been hit by five bombs. One struck the forecastle, exploded on the main armory deck, backfired and blew up forward of No. 1 turret gun, said Johnson. A direct hit by an incendiary bomb melted and destroyed the entire bridge structure.

Fires were raging below deck. The battleship was sinking by the bow. Casualties would include 50 officers and crew killed and at least 109 wounded.

Miraculously, the *Nevada* ran the gauntlet, but was immediately confronted by a new peril. Listing badly and its navigational equipment mostly demolished, the *Nevada* would not be able to make the turn that would take it out of Pearl Harbor and into open sea. If the *Nevada* went down in that narrow channel, it could become not only a navigational hazard blocking the entrance to the Pacific Fleet's headquarters base, but an easier target. The only alternative was to beach her which was done shortly after 9 a.m., at Landing C, Hospital Point, where many of her casualties could be put ashore.

The *Hoga*, a quarter of a mile north of the *Nevada* and just clearing 1010 Dock, saw the battleship's plight. McManus rang for full speed ahead as the ferocity of the Japanese attacks continued. Glancing up, McManus saw a triangular pattern of bombs hurtling down dead ahead of the *Hoga*. He put the boat full astern as the missiles plopped into the channel, but failed to explode. Still the Japanese refused to be denied the *Nevada*.

As the *Hoga* closed upon the battleship, a bomb believed to have been intended for the *Nevada*, struck the forward deck of the destroyer *USS Shaw* in Floating Drydock No. 2. The *Shaw*'s ammunition stores exploded; wreathing the destroyer in flames and smoke crowned by flaming trails of skyrocketing shells.

By 9:25, according to the *Hoga*'s journal, she had put her port bow alongside the port bow of the grounded *Nevada*, whose on-board firefighting system had been blasted away. The *Hoga* switched to firefighting mode. A stream jetted from the monitor on the pilothouse into the flames while *Hoga* crewmen coupled four hoses to its discharge outlets. The *Nevada*'s crew pulled the hoselines on board and attacked the fires. Intense smoke and flames below deck involved storerooms and bedding which coughed flame and smoke from bomb holes and open hatches in the forward section. The *Nevada*'s hull rapidly heated until layers of gray paint on the hull began burning.

Still the fires worsened. The *Nevada*'s crew called for more hose. The *Hoga* had none to give. "Better start thinking about forming a bucket brigade," McManus called to them. With the *Hoga* secured to the *Nevada*, McManus watched with growing worry as the anchor continued to settle until it was directly opposite him. The *Hoga*'s pilothouse and top-mounted monitor gun soon stood well above the *Nevada*'s bow. There was no question that the battleship was going down at the strategically worst possible place: the entrance to Pearl Harbor.

A clever plan was devised, recalls Johnson. The *Hoga* would help push the *Nevada* backwards across the channel where it would be beached, stern first. That would not only clear the sea lane, but point the *Nevada*'s guns out to sea. As a beached fortress, the *Nevada* would serve as a deadly warning to Japanese submarines and other warships reported to be preparing to invade Pearl Harbor. Unknown to the enemy, of course, was the fact that the guns of the *Nevada* were useless because of the heavy damage she had suffered.

The tiny *Hoga*, dwarfed by the *Nevada* which was nearly six times longer than the tug, began pushing the battle-

The **Hoga,** a U.S. Navy combination yard tug and fireboat, rushed to the aid of the *USS Nevada* after the battleship was severely damaged and set afire during the Pearl Harbor attack. Mooring to the port bow of the **Nevada,** the **Hoga's** pilothouse monitor gun bored into the flames while the tug's fire pumps fed

hoselines put aboard the battleship. The **Hoga** pushed the **Nevada** across the channel leading to open sea and helped to beach her at Waipio Point to avoid the threat that the sinking **Nevada** would block the entrance to Pearl Harbor. **USS Arizona** Memorial, National Park Service

wagon across the channel while its pilothouse monitor gun continued to bore into the flames. The cross-channel maneuver was a navigational feat that would long be remembered because it was photographed for posterity by a Navy photographer.

With the *Nevada's* engines reportedly at one-third speed astern and the *Hoga* pushing full ahead, the channel crossing seemed to take forever. Just before the *Nevada* was grounded near a Waipio Point sugar cane field, her engines were stopped. The *Hoga* gave a mighty push and the battleship's stern went solidly aground as the anchor was dropped. The *Nevada* was no longer a navigational threat to other vessels.

Firefighting continued. At 12:40, according to her log, the seaplane tender *Avocet* arrived to lend more firefighting assistance. Fires were not controlled until 2:30 p.m. With the *Avocet* continuing to help, the greater firefighting capabilities of the *Hoga* were needed along Battleship Row. The *Hoga's* log shows she first attacked flames aboard the *West Virginia* and other warships along the row starting at 1:30 p.m. Three hours later the *Hoga's* streams began boring into flames raging inside the blackened *Arizona.* The *Hoga* remained at the *Arizona's* side for two days. The fire tugboat's crew saw, but could not help, much less recover, many of the battleship's dead and drowned officers and crew.

At 1 p.m., December 9, two days after the sneak attack, the *Hoga* was relieved of its firefighting duties and returned to its berth at Yard Craft Dock to await further orders. The following days, according to the *Hoga's* log, included search for bodies, removal of damaged ships, and debris from the channel and patrol searches for reported enemy submarines.

Almost constantly in the line of fire during the two-hour attack, none of the *Hoga's* crew was injured, nor was the boat damaged. Somehow the story started and continued through 1988 that a large dent on the front port quarter of the *Hoga* is a memento of work while pushing the *Nevada* onto a spot which today is known as Nevada Point. In 1988, McManus and Brown continued to insist that the crease in the hull never occurred during their time on the *Hoga.*

The *Nevada* was raised and refloated, February 12, 1942, several months after the surprise attack. Following a $23 million reconstruction and modernization at the Puget Sound Navy Yard, Bremerton, Wa., she rejoined the fleet. The *Nevada* saw much World War II action, including the invasions of France, Iwo Jima and Okinawa. Later, she had the ironic experience of participating in the occupation of Japan when the war ended. The *Nevada* was used as an experimental vessel to test radiation during the June, 1946, atomic bomb tests at Bikini. Decommissioned that year, the

*Nevada* was used as a target by other warships testing new armament. At 2:34 p.m., July 31, 1948, gunfire and aerial torpedoes sent her to the bottom of the ocean off Hawaii.

The *Arizona*, still laying where she sank, was dedicated on Memorial Day, 1962, as the *USS Arizona* Memorial. The more than 1100 Navy and Marine Corps personnel who went down with her are forever entombed in her hull. A total of 2403 military personnel and civilians were killed in the surprise attack. Bombardier Kanai, whose mammoth bomb sunk the *Arizona*, was shot down and killed during a Wake Island battle shortly after Pearl Harbor.

Not until years later did McManus and Brown learn of the *Hoga* citation by Admiral Chester W. Nimitz. As commander-in-chief of the Pacific Fleet, Nimitz officially recognized the *Hoga's* crew "for distinguished service in the line of your profession ... efficient action and disregard of your own personal safety ... when another ship (*Nevada*) was disabled and appeared to be out of control, with serious fires in the forepart of that ship, you moored your tug to her bow and assisted materially in the beaching operations in an outstanding manner. Furthermore, each member of the crew of the *Hoga* functioned in a most efficient manner and exhibited commendable disregard of personal danger throughout the operations."

Seven years after the Pearl Harbor attack, the *Hoga* began a new mission when it was loaned by the Navy to the Port of Oakland, Ca., to serve as that city's first fireboat. Following the official transfer of papers, May 28, 1948, more than $73,000 was spent by Oakland to increase the boat's capacity from 4000 to 10,000 gpm at 150 psi to meet the needs of the rapidly-growing port. Upgrading was done by the Pacific Coast Engineering Company at Pacific Drydock and Repair in Oakland.

Retained were the *Hoga's* twin McIntosh & Seymour (American Locomotive ) of Auburn N.Y., 650 horsepower at 740 rpm main diesel engines which operate two 419 kilowatt Westinghouse electric generators for propulsion or pumping requirements. The boat's pumping capacity was increased by the addition of three United States Iron Works pumps, each single-stage centrifugals. They were powered by three 250 horsepower supercharged Buda diesels. One of these was later replaced by a D333 Caterpillar.

Because there was insufficient engine room space for these additional pumps, they were installed on the main deck abaft the deckhouse. This above and below deck pump location is believed to be unique among fireboats. The weight of these aft-mounted pumps largely resulted in a new draft for the boat: 9-feet forward and 12-feet aft. As rebuilt, the boat weighs 343 net tons.

Also part of the upgrading was the installation of additional turret guns. Three were located port, starboard and center at the end of the upper deck platform. The fourth was mounted at the stern. The original tower-mounted gun is elevated about 14 feet above the upper deck. Another gun is mounted on the pilothouse lip.

In 1989, the *City of Oakland* had six monitor guns. The guns and 12 gates hose discharge outlets, evenly distributed on the port and starboard sides of the deckhouse, are supplied by a common loop, or main. This loop encircles the upper area of the deckhouse. A series of fog nozzles, mounted immediately above the loop and supplied by it, provide protective sprays against radiant heat. By 1989, four foam-discharging hose outlets were fitted under the pilothouse lip. The boat carries 200 gallons of high expansion foam.

*The **Hoga** was loaned by the U.S. Navy to the Port of Oakland after World War II. The combination yard tug and fireboat was converted into a single-purpose fireboat. While the boat retains most of its original configuration, its pumping capacity was raised from 3000 to 10,000 gpm at 150 psi. Note the unique outboard water supply main that encircles the deckhouse. The original pilothouse monitor which was used to fight fires resulting from the Pearl Harbor attack, remains in use, but the four hose discharge outlets on the deckhouse do not. David C. Henley*

*Oakland's fireboat is believed to be unique because of its above and below deck pumps. There was insufficient space in the engine room for more pumps when the U.S Navy loaned the **Hoga** to Oakland. To boost the former Pearl Harbor tug's pumping capacity from 3000 to 10,000 gpm, three more pumps were aft-mounted on the main deck to give the vessel a total of five. From the Collection of Bill Dahlquist*

*When Pearl Harbor's **Hoga** was loaned to Oakland in 1948, the city increased the boat's pumping capacity to 10,000 gpm at 150 psi. On December 2, 1948, the day after the boat went in service, it was called to flood the hold of the burning **Hawaiian Rancher**.* David C. Henley

Renamed the *Port of Oakland* (changed to *City of Oakland* after 1955) the fireboat went in service December 1, 1948. And none too soon. The following day the fireboat met the *Hawaiian Rancher,* a Matson freighter inbound from Stockton, Ca. The captain had reported a fire in No. 2 hold as he came into the San Francisco Bay and approached the San Francisco-Oakland (Transbay) Bridge.

The *Port of Oakland* escorted the ship into the Encinal Terminal, Alameda. Intense heat and smoke forced flooding of the hold with hoselines supplied by the *Port of Oakland.* The fireboat narrowly escaped severe damage or destruction when the *Hawaiian Rancher* suddenly listed and seemed about to capsize. A year later the *Port of Oakland* pumped for more than nine hours as it delivered more than six million gallons of water while battling the Pier 4, Oakland Army Base, Outer Harbor, fire, March 15, 1949. The fire was, at that time, the largest and most destructive in the history of the Oakland Fire Department.

Considering her historic fame, the *City of Oakland* was the obvious choice to take President Jimmy Carter on a July 3, 1980, tour of the port. The crew presented him with a copy of the famous photo of the boat fighting fires aboard the *Nevada.* President Carter pointed a *City of Oakland*'s monitor gun at a boat carrying newsmen and photographers. It is said that several fireboat crewmen encouraged the president to open the discharge valve.

Commemorating Pearl Harbor Day, the *City of Oakland* was moved to a new berth and fire station, December 7, 1982, at the foot of Clay Street, two blocks west of Jack London Square. Its operational costs are shared by the Port of Oakland and the Oakland Fire Department. Oakland firefighters crew the boat. Veteran of many San Francisco Bay and Golden Gate area fires, the *City of Oakland* is the only remaining active vessel afloat among those at Pearl Harbor during the 1941 attack.

Late in 1988 discussions were continuing among officials of Oakland, the U.S. Navy and the National Park Service. If all goes as hoped, the historical landmark, *City of Oakland,* ne *Hoga,* will soon return to Pearl Harbor to become an everlasting part of the *USS Arizona* Memorial.

The *Hoga* was not alone among fireboats to earn historic honors during World War II. More than a year before Pearl Harbor, the war in Europe was going badly for the Allies. Belgium was surrendering and France was about to fall. German armies forced the retreat of several hundred thousand troops, mostly British, to the beaches and harbor at Dunkirk, about 50 miles across the channel from England. The evacuation of Dunkirk, code-named Operation Dynamo, involved hundreds of British and French warships with support from the entire British Metropolitan Air Force.

Much of the credit for the successful evacuation of more than 300,000 British and French troops went to what Winston Churchill called The Mosquito Armada of nearly 400 small craft, including tugs, trawlers, motorboats and yachts. They set out from England to help rescue troops, many of them badly wounded and unable to walk. The Mosquito Armada of little ships, mostly crewed by civilian volunteers, was essential because their shallow draft enabled them to get close to the harbor and beaches. Dunkirk was aflame as German artillery and aircraft laid down incessant attacks to wipe out those who were trapped and those who would try to save them. Allied warships fought back, but more than 240 of them were sunk as were 182 boats of The Mosquito Armada.

In The Mosquito Armada was London's five-year-old fireboat, *Massey Shaw*, named after Sir Eyre Massey Shaw, from 1866-1891, the first chief of the London Metropolitan Fire Brigade and regarded as the father of London's fireboat fleet because of his interest and innovations.

The *Massey Shaw* was as ill-equipped for Operation Dynamo as the *Hoga* was for the Pearl Harbor attack. Built at the J. Samuel White Shipyard at Cowes on the Isle of Wight and launched February 25, 1935, the 50-ton steam-turbine *Massey Shaw* is steel-hulled with two propellers driven by a pair of 160 horsepower Glennfer diesel engines at a top speed rated at 12 knots. The *Massey Shaw's* diesels could be switched to operate twin four-stage centrifugal pumps which delivered 3000 gpm. Her major firefighting feature was a large solid copper and brass monitor gun mounted on the main deck abaft the pilothouse. With its 3½-inch-diameter nozzle, the *Massey Shaw* could throw around 750 tons of water an hour.

*The **City of Oakland's** guns "cool" the hull of the U.S. Navy Hospital Ship, **Mercy**, during a November, 1987, training exercise by members of the U.S. Coast Guard Reserve as well as Oakland and San Francisco firefighters. Also participating in the drill was San Francisco's **Phoenix** (upper right). Gary Pirkig*

Typical of London fireboats, the *Massey Shaw's* low profile enabled it to pass under bridges. Measuring 78-feet-long and 13½-feet in beam, it has a draft of only 3-feet, 9-inches. It was this latter figure that made her ideal for fighting fires along the Thames River and its shallow inlets and perfect for operating in Dunkirk's shallow water.

Conversely, the *Massey Shaw* was an extremely dangerous craft to use while crossing the invariably turbulent channel. The boat was primarily chosen for Dunkirk as a firefighter; a mission that proved impossible when the armada was confronted with mass pandemonium along the beaches as thousands of troops were pinned down by enemy fire or were trying to wade into the surf where they could be rescued.

On Thursday afternoon, May 30, 1940, the British Admiralty called for volunteers to crew the *Massey Shaw* for the possibly fatal undertaking. There were no lack of them. Thirteen were selected under command of Aubrey John May, a London Fire Brigade sub-officer: four firemen, a sub-officer and a station officer of the brigade and six Auxiliary Fire Service firefighters.

The volunteers quickly put aboard provisions and firefighting foam. Someone hurried to a store to buy a cheap compass because the *Massey Shaw* had none. Up to that day there had been no need of one while fighting fires along London's waterfront. The compass proved the adage that you get what you pay for. The instrument proved of little value.

Shortly after 4 p.m., two hours after they had been selected, the *Massey Shaw's* crew got underway. As the fireboat steamed to its rendezvous with other little ships at Ramsgate, her crew slathered camouflage gray paint over the glistening monitor gun, ship's fittings and the white cabin. Timbers were brought aboard at Ramsgate and spiked to the *Massey Shaw's* windows to protect against breakage and swamping of the fireboat by anticipated high seas.

After an overnight stay to bring the armada up to full strength and taking on board a Royal Naval Officer as navigator, the armada set out across the channel. With its bow scooping into the sea troughs, the *Massey Shaw* rolled dangerously from port to starboard. Around 3 p.m., May 31, the *Massey Shaw* was within three miles of Dunkirk where her crew saw the glare of the burning town, the pall of smoke shrouding it and heard the explosions of shells raining upon Dunkirk and its beaches.

Arriving off Dunkirk, the *Massey Shaw's* crew could only liken the death and destruction along the beaches and the offshore waters to a graveyard of ships. Maneuvering among them in heavy surf to get close enough to pick up survivors was treacherous. As shallow-drafted as the *Massey Shaw* was, there was constant fear of running aground on a shoal, fouling its propellers in the flotsam and jetsam, hitting a mine or taking a direct hit from the shelling. Constantly bumping and ramming other boats in the armada, the *Massey Shaw* suffered only minor damage as its crew dragged and lifted soldiers on board. Some were about to drown. Several *Massey Shaw* firemen jumped overboard and saved them.

The *Massey Shaw* made three round trips from Dunkirk to Ramsgate. On the first they had taken on board 65 British

The London Fire Brigade's **Massey Shaw** was the only fireboat among what
Winston Churchill called The Mosquito Armada of little ships which participated
in the World War II evacuation of troops from the harbor and beaches around
Dunkirk. The **Massey Shaw**'s all-volunteer crew was credited with rescuing 646
military personnel, including 37 survivors of a French freighter which sank after
hitting a mine. **Massey Shaw** and Marine Vessels Preservation Society, Ltd.

soldiers, 30 of whom were crammed into a cabin built for
six. Many were seasick as the *Massey Shaw* headed to
Ramsgate with their load that night. The fireboat's position
was betrayed by its twin wakes which formed phosphores-
cent-like streams in the moonlight. A German plane spotted the *Massey Shaw*, swooped low and dropped a bomb.
The fireboat was not damaged, nor was anyone aboard hurt
by the jolting near-hit.

Landing the troops at Ramsgate at 7 a.m., June 1, the
*Massey Shaw*'s crew took a brief respite, refueled and took
on board a Lewis machinegun which they would not fire.
But it gave them some comfort to know they had it, even if
they were too busy to use it. Returning to Dunkirk that
night, the glare of blazing Dunkirk lit the beaches as German shells hit closer. The next hours were a blur of rescues,
near groundings and collisions with other craft in the
armada; their crews working as hard and as desperately as
the *Massey Shaw*'s to load up and get out to sea.

The *Massey Shaw* brought 106 troops to safety at
Ramsgate. And she picked up five loads of soldiers, many
of them wounded and on stretchers, and off-loaded them
on warships standing a mile out from Dunkirk. British

Navy guns pounded German land installations and
directed machine gun fire at attacking planes.

The *Massey Shaw*'s shallow draft enabled her to be one of
the most effective of The Mosquito Armada and was the last
to leave its sector when the evacuation efforts successfully
ended. As she prepared to go home, the fireboat's anchor
and its chain snagged an underwater object and were lost.
Leaving Dunkirk at 3:30 a.m., June 2, the fireboat arrived
five hours later at Ramsgate with around 40 more sailors.
Altogether the *Massey Shaw* was credited with evacuating
more than 600 soldiers.

The mission done, the *Massey Shaw* was ordered to return
to London. Her crew, marveling at their incredible fortune
in not losing a man or the fireboat, congratulated each other
on a job well done. But it was not the end of the *Massey
Shaw*'s saga of Dunkirk. Off the coast of Margate they spotted heads bobbing in oil-slickened waters. These were the
survivors of the French freighter, *Emil de Champ*, which had
struck a mine and sunk. The *Massey Shaw* steamed full
ahead into the oil-slimed waters and saved 37 sailors, many
of them maimed, and took them into Ramsgate.

*Detroit's **James Battle** was built around 1900 with coal-fired boilers and reciprocating piston pumps. Dieselized in 1936, the fireboat's capacity was increased to 7200 gpm at 150 psi. During World War II, the **Battle** was taken out of retirement to serve as an Allied Nations convoy vessel operating out of Halifax, Nova Scotia. After the war, the **Battle** did dual duty as a Montreal fireboat and as a tugboat on the St. Lawrence Seaway. From the Collection of Steven Lang*

Built to fight fires, not wars, the *Massey Shaw*, everlastingly famous for rescuing 646 soldiers and sailors, headed for home. Londoners were alerted to the imminent arrival of the *Massey Shaw* and the feats of her volunteer crew.

As the *Massey Shaw* steamed up the Thames and her berth at Blackfriars Bridge at the Lambeth Headquarters of the London Fire Brigade, fireboat crews cheered her as she passed. A reception at Lambeth was hosted by Major Frank Whitford Jackson, commander of the London Fire Brigade and the Auxiliary Fire Services. Also on hand to welcome and congratulate the *Massey Shaw*'s crew were the volunteers' wives, mothers and friends, as well as ranking officers of the brigade.

The *Massey Shaw* and her crew won many honors. The fireboat was the only civilian-manned craft in the armada to be cited by Vice Admiral Sir Bertram Ramsey in his dispatches to the Lords Commissioners of the Navy. With the hearty approval of His Majesty, King George VI, and the Admiralty Board, Sub-Officer May was awarded the Naval Distinguished Service Medal. The medal, traditionally given in recognition of the service of Royal Navy and Royal Marines, had never been awarded to a fireman. Two other *Massey Shaw* crewmen, Henry Albert William Ray and Edmund Gordon Wright, both of the Auxiliary Fire Services, were cited in naval dispatches for their heroism.

The *Massey Shaw* saw much fire duty during World War II, including the Sunday night blitz, December 29, 1940, when the fireboat and 2000 other pieces of apparatus fought 1500 fires touched off by incendiary bombs. The fireboat battled many other London Blitz incendiary fires along the waterfront and supplied water to as many as four land-

*Most of America's fireboats were aging at the start of World War II, but cities could not replace them because shipyards were working around-the-clock building warships. Boston's **John P. Dowd** did its part for the war effort until it was retired in 1943 after 34 years of service. The **Dowd** had a 6000 gpm at 150 psi pumping capacity and weighed 179 tons. It was 113-feet-long, with a 26-foot beam and a 9-foot draft. Built in 1909 at a cost of $92,625 by Bertelson & Peterson, the **Dowd** was designed by William F. Keough. From the Collection of Bill Noonan*

No less vital to the war effort were inland ports, such as Chicago, where the **Fred A. Busse,** one of the last fireboats built before the outbreak of World War II, helped to protect the flow of war materials to seaports. The **Busse** was rated at 10,000 gpm at 150 psi by the National Board of Fire Underwriters. This photo prompts the author of **Fireboats** to recall the postwar 5-11 alarm lumber yard fire along the Chicago River during a snowstorm. As a reporter, he boarded the moored, ice-caked **Busse** for a hot cup of coffee. From the Collection of Paul Ditzel

based engines when bombs destroyed mains. Four hundred and fourteen men and women of the London Fire Brigade and the Auxiliary Fire Services were killed during World War II while fighting fires which threatened to destroy London.

At the age of 36, the *Massey Shaw* was taken out of service in 1971. The fireboat, now an historic landmark and tourist attraction, is berthed at London's East India Docks at the South Quay Waterside Development. The *Massey Shaw* is close to restoration by the *Massey Shaw* and Marine Vessels Preservation Society, Ltd., a non-profit organization.

The United States' declaration of war in 1941 immediately put heavy burdens upon the nation's seaports. Ship traffic surged with troop embarkations and war materials handling. More cargoes and more ships meant potentially more fires, especially with the large volume of petroleum products, ammunition and other explosives. Sabotage was a constant fear as was enemy air attack and even invasion. If Pearl Harbor could be devastated by surprise, why not Los Angeles, New York, or other hubs for feeding war supply lines? Most fireboats along the Atlantic, Pacific and Gulf Coasts were repainted camouflage gray.

With shipyards working around the clock for the war effort, cities were forced to put aside thoughts of new fireboats, badly as they might have been needed. New

Orleans' 10,000 gpm at 150 psi *John H. McGiven,* put in service in 1942, was the only large fireboat built for an American city during the 1941-1945 war years.

Wartime experience showed the fire service had more to fear from carelessness in ports than a saboteur's bomb. Peacetime fire safety regulations were often ignored. The 1942 loss of the French ocean liner, *Normandie,* was a prime example. This world's largest luxury liner was being converted into a troopship at New York's Pier 88 on the North River at the foot of West 45th Street. A welder's torch is believed to have ignited kapok-filled life preservers. Instead of immediately turning in an alarm, workmen fought the rapidly worsening fire for at least a quarter of an hour.

At 2:49 p.m., February 9, 1942, the first of five alarms was sounded from Box 852 for the burning *Normandie.* Fireboats and hundreds of firefighters from land-based companies poured millions of gallons of water in the liner as fire gnawed deeper in the *Normandie.* The fight to save the $50 million liner continued through the night. The following day the *Normandie* slowly rolled over on her port side while the *Fire Fighter,* refusing to quit, continued to pour water into a ship that was lost.

Germany celebrated what their propagandists claimed was the successful sabotage of the *Normandie.* The National Fire Protection Association's report proved otherwise.

Although the $50 million **Normandie**, the world's largest luxury liner at the time, lays capsized and ruined, New York's fireboat **Fire Fighter** refuses to quit the battle to save her. The **Normandie** was being converted into a troop carrier when a welder's torch is believed to have ignited a fire on February 9, 1942. Before calling the fire department, workers and supervisors fought the worsening fire with a scenario which the National Fire Protection Association described as "a Hollywood slapstick comedy." When the alarm was finally turned in, hundreds of firefighters from land and fireboat companies were unable to prevent the loss of the **Normandie** to the war effort. The liner capsized the following day. From the Collection of Paul Ditzel

Fireman Joseph V. "Rocky" Roquemore, alone aboard Los Angeles' **Fireboat No. 3**, single-handedly rescued 17 sailors and workers who had jumped or were thrown into the water when explosions and fires erupted from Navy landing craft and a wooden dock on October 21, 1944. The $10,000 boat, commissioned March 13, 1928, was wooden and gasoline-powered; further adding to the dangers Roquemore faced during his rescues which remain unparalleled in the history of the Los Angeles fireboat fleet. The boat, mainly intended as an auxiliary to **Fireboat No. 2**, was 38-feet-long, with a beam of 10-feet and was shallow-drafted. Its 275 horsepower engine served to propel and operate the boat's 290 gpm pump. The boat was decommissioned, July 28, 1967, and sold. From the Collection of Bill Dahlquist

Sharply critical of the negligence and ineptitude of the ship's workers and their supervisors, the NFPA report said their efforts had "the elements of a Hollywood slapstick comedy" before they thought to call the fire department.

A year later, the Port of New York area was threatened by catastrophe when fire broke out after a fuel pipe burst in the boiler room of the Panamanian freighter, *El Estero*, berthed at the U.S. Army Caven Point Terminal in Bayonne, south of Jersey City. The *El Estero* had 1400 tons of high explosives on board and was loading high-octane aviation gasoline when the fire was discovered.

Perhaps remembering the *Normandie*, an alarm was immediately turned in at 5:20 that Saturday afternoon, April 24, 1943. The deep-seated fire quickly spread toward the ammunition cargo holds and gasoline compartments of the *El Estero*, berthed in an area studded with clusters of storage tanks containing millions of gallons of petroleum products.

Despite efforts of Jersey City firefighters, the situation was worsening an hour after the first alarm, even with New York's two largest fireboats, the *Fire Fighter* and the *John J. Harvey*, together deluging 27,000 gallons of water a minute into the smoking and rapidly heating *El Estero*. Port authorities, fearing a catastrophic explosion which would cause devastation throughout the Port of New York area, ordered the *El Estero* hauled from Cave Point to what was hoped would be a relatively safe area in open sea where she could blow up with relatively little damage.

Radio stations interrupted programs to warn listeners to cover or stay away from their windows and to prepare for possible building collapses. If the *El Estero* exploded—and at that point everyone was certain she would—the devastation would be widespread, including the Statue of Liberty, less than a mile east of the *El Estero*.

Under tow by 6:20 p.m., by Coast Guard craft and tugs, the *El Estero* started toward The Narrows leading to open sea. The *Fire Fighter* and the *John J. Harvey*, tied up alongside the freighter's port and starboard sides, continued to cascade tens of thousands of gallons into her. Under the weight

New York's fireboat **John J. Harvey** hurried to the aid of the burning **El Estero**, loaded with ammunition and gasoline. With a disastrous explosion expected momentarily that Saturday afternoon, April 24, 1943, the 17,000 gpm **Harvey** and the 20,000 gpm **Fire Fighter** cascaded water into the **El Estero** as she was being hauled to sea. The weight of the fireboats' water finally caused the freighter to sink off Robbins Reef Lighthouse. An almost certain catastrophic explosion which would have knocked out much of the vital New York port area was avoided. Port Chicago, CA., was not as lucky, July 17, 1944, when two ammunition ships exploded and devastated much of the port. Killed were 322 people in what was World War II's worst waterfront disaster in the United States. The cause was never officially determined. From the Collection of Bill Noonan

The **Fire Fighter**, flagship of the Marine Division, Fire Department of New York, was one of two fireboats and crews cited for heroism when they flooded the **El Estero**, loaded with high explosives and high-octane aviation fuel. Fire was discovered aboard the freighter, April 24, 1943, when it was being loaded at Bayonne, N.J. Fearing a catastrophic explosion, port authorities ordered her towed to sea. Along with the FDNY's second most powerful boat, the 17,000 gpm **James J. Harvey**, the fireboats pumped 27,000 gallons of water a minute into the **El Estero** while she was being towed to sea where she sank. The **Fire Fighter's** tower, normally resting in a horizontal cradle, is shown being raised. From the Collection of Bill Noonan

of the water sloshing inside her, the *El Estero* suddenly listed to starboard. Cargo and parts of the ship crashed onto the *Fire Fighter*. Its crew, expecting the *El Estero* was about to capsize, cast off. When it did not, the *Fire Fighter* quickly returned and opened up its deck pipes again. If fireboat streams could not penetrate the heart of the fire, perhaps they could flood the cargo holds and the ship would go down before she exploded.

Around 9 p.m., the *El Estero* was about one-half mile due west of Robbins Reef Lighthouse when the freighter sank. The disaster alert was lifted. New York fireboat firefighters and Jersey City firemen were awarded citations for their heroism in averting an almost certain catastrophe.

Bravery was commonplace among fireboat firefighters in Los Angeles during World War II where gross negligence touched off a disastrous explosion and fire. The tankship *Fredericksburg* was at the Union Oil Terminal's Berth 151 that Saturday afternoon, October 21, 1944, and taking on a cargo of toluene, an anti-knock additive to military aircraft fuel. Ship security, if indeed there was any, failed to notice that toluene was leaking from the tanker.

The leak went undetected for at least six hours as a volatile smear fanned out from the *Fredericksburg*. The tide carried this floating bomb across the Turning Basin to Berth 223, Terminal Island. A blanket of toluene surrounded two Navy landing craft moored side by side, while welders prepared them for war. Still seeking ignition, toluene vapors accumulated under a 200-foot-long wooden wharf where Navy and civilian workers had parked their cars and trucks.

Ignition, probably caused by a welder's torch, was followed by sheets of flame and a towering cloud of black smoke. The 1:58 p.m., explosion was seen and heard by firefighters at *Fireboat No. 2*, berthed a few hundred yards south. The fireboat and its auxiliary, *Fireboat No. 3* arrived in minutes.

As Pilot Brainard "Choppy" Gray brought *Fireboat No. 2* close, Captain Maurice P. Allen ordered the fireboat's most powerful gun, Big Bertha, to attack. Opening the discharge valve, a fireman directed a 10,200 gpm wallop. With one sweep, Big Bertha knocked down flames aboard both landing craft. A second sweep successfully battered flames on the dock, its crane and 25 vehicles. Big Bertha's punch pushed a large truck completely across the wharf.

Incredible as Big Bertha's effectiveness proved that afternoon, the hero of the day was Senior Operator Joseph V. "Rocky" Roquemore, alone in *Fireboat No. 3*. The 16-year-old wooden boat was not much of a firefighter. Only 38-feet-long and driven by a 275 horsepower gasoline engine, its pump could put out but 290 gpm.

Spotting many sailors and workers who had been blown or jumped from the dock and landing craft, Roquemore gave full throttle and steered the old boat close to the radiant heat and residual fires around and under the wharf. Tossing life preservers to the men in the water, he began pulling survivors on board. Many were badly burned or suffered other injuries.

Quickly putting them ashore where medical aid awaited, Roquemore sped back to rescue more sailors and workers in imminent danger of drowning. He was credited with saving 17 lives. For lack of safety precautions aboard the *Fredericksburg*, 11 sailors and five workers died. Thirty-five others were injured.

Roquemore's feat remains unparalleled in the history of the Los Angeles fireboat fleet. Although the Los Angeles Fire Department has medals for such bravery, Roquemore's heroism was never officially recognized.

Fire Chief John H. Alderson explained: "Rocky was only doing what firemen are paid to do and what duty compels them to do."

*National security often caused authorities to withhold details of wartime fires, especially those in defense industries. Few details were released about this May 30, 1945, fire among U.S. Navy warships at Los Angeles Shipbuilding and Drydock Corporation (Todd Shipbuilding) at Berths 103-110 in San Pedro. Army and Navy tugs, along with Los Angeles **Fireboat No. 2** (note water tower, left of center) joined in the battle. From the Collection of Bill Dahlquist*

*San Francisco's 4000 gpm fireboat, **Dennis T. Sullivan,** helps to battle a Pier 48 fire which is believed to have occurred during World War II. The **Sullivan** was named after San Francisco's fire chief who was fatally injured by the first jolt of the Great Earthquake of 1906. W.C. Dunn*

The **Mistress Delta** fireboat of Stockton, CA, moves in to fight a second-alarm fire, October 20, 1964, involving a dock and a wooden boat. The **Mistress Delta**, originally built in the U.S. Navy Yard, Alameda, for the Coast Guard, was declared surplus after World War II. The Port of Stockton bought the boat for $500 and spent around $3000 to turn it into a 2000 gpm at 150 psi fireboat. The 40-foot-long **Mistress Delta** had a beam of 14½-feet and a draft of slightly more than 4-feet. It served until April 13, 1977. SFD

Boston's **Joseph J. Luna** fireboat was named in memory of Fireman Joseph J. Luna of Ladder 9, who was killed in action during World War II while serving with the U.S. Marine Corps. The **Luna,** formerly the U.S. Navy minesweeper, **USS Marabout,** was a sister ship to the **USS Bulwark,** also purchased by Boston as war surplus. The **Luna** and the **James F. McTighe** were each rated at 6000 gpm at 150 psi capacity. They went in service in 1947 and lasted 25 years. They were the first new fireboats in the Boston fleet since the **Matthew J. Boyle** was commissioned 16 years earlier. From the Collection of Bill Noonan

Although it had no firefighting capabilities, **Smoke I** of the Fire Department of New York's Marine Division, saw plenty of action in the port during World War II. The launch speeded officers of the Marine Division to waterfront fires where the boat was used as a command post as well as a shuttle carrier of additional personnel to the fireboats. Commissioned November 16, 1938, and costing $26,000, the 17-knot launch was powered by a Sterling gasoline engine. The 55-foot-long wooden boat was taken out of service April 16, 1958. From the Collection of Bill Noonan

The **William J. Brennan** was one of two World War II surplus U.S. Coast Guard boats acquired by Newark, N.J., after World War II and converted into fireboats. The **Brennan** and its sister, **Michael P. Duffy,** each had a pumping capacity of 2000 gpm at 150 psi. The boats were 40-feet-long. From the Collection of Bill Noonan

When small boats were built to support the war effort, government officials did not foresee the time when many of them would be declared surplus. Cities large and small eagerly purchased these craft for inexpensive conversion into fireboats. The Boston Fire Department purchased the U.S. Navy's coastal minesweeper, **USS Bulwark**, for $25,000. An additional $168,000 was spent rebuilding the minesweep into the fireboat **James F. McTighe** at the General Ship & Engine`

Works, Boston. Put in service December 31, 1947, the **McTighe** had a capacity of 6000 gpm at 150 psi and was a welcome addition to Boston's aging fireboat fleet. The diesel-powered fireboat was 97-feet-long, with a 21-foot beam and a draft of 9-feet. The 10-knot fireboat saw service for a quarter of a century before retirement, June 6, 1972. From the Collection of Bill Noonan

# CHAPTER SEVEN
## BLANKET OF FLAMES

The worst waterfront catastrophe in United States history occurred April 16, 1947, when the ammonium nitrate laden freighter, **Grandcamp**, burned and exploded at its Texas City dock. Ensuing explosions and fires killed 561 people. Sixteen hours after the **Grandcamp** blew up, another burning freighter, the **High Flyer**, loaded with sulphur, also exploded. Fires burned along the waterfront and in the harbor for days. Authorities asked: Would fireboat protection have prevented the disaster, or at least minimized damage? From the Collection of Paul Ditzel

World War II and its aftermath marked the start of escalating hazardous materials problems and other fire protection concerns, along with the growth of ports which had not earlier needed fireboat coverage. That the age of hazardous materials had arrived was catastrophically demonstrated April 16, 1947, when the French freighter, *Grandcamp*, loaded with 2300 tons of ammonium nitrate, burned and exploded at its Texas City pier, 10 miles north of Galveston.

The first explosion at 9:12 a.m., and those which quickly followed, were felt for 150 miles. Seismograph needles jiggled at Denver's Regis College earthquake center. The explosions and fires devastated an area 20 blocks long and 12 blocks wide, including ships, warehouses, oil refineries, tank farms, the Monsanto Chemical Company and surrounding commercial and residential districts. Board Chairman Edgar M. Queeny of Monsanto said the *Grandcamp's* initial blast was equal to the simultaneous detonation of 250 World War II five-ton blockbuster bombs.

The waterfront catastrophe, worst in American history, killed 561 people, injured 3000 and caused $50 million in damages. Among those killed was Fire Chief Henry. J. Baumgartner and 26 members of the Texas City Volunteer Fire Department. While fires burned, experts hurried to Texas City to determine if their ports were safe from similar disasters. One question asked: If Texas City had fireboat protection, would damage have been minimized? The *Grandcamp* fire was fought by land companies for an hour before the explosion. The *High Flyer*, loaded with 2000 tons of sulphur, burned for 16 hours and blew up. Fires along the waterfront and among ships in the harbor burned unchecked for days.

Among those who left immediately for Texas City were three officers of the Los Angeles Fire Department. "We must all think and prepare for these things," said Battalion Chief Harry Gross upon his return. "Can it happen in Los Angeles? It did happen in Texas City." Six weeks after the disaster and Gross's prophetic warning, a waterfront calamity erupted in the Port of Los Angeles and proved once again that only one fire can justify the cost and upkeep of fireboats.

*Only the bow of the **Markay** is visible after the petroleum tankship suddenly exploded in Los Angeles, June 22, 1947, six weeks after the Texas City disaster. This firefighting challenge confronted firemen when they were jolted awake by the explosion and pulled out of their fireboat house several hundred yards downstream. The fire had already involved the Shell Oil Terminal wharf (right) and the American President Lines wharf and warehouses (left). From the Collection of Paul Ditzel*

*Los Angeles' **Fireboat No. 2** prevented the spread of flames to a heavily-industrialized and residential area after what was a potentially suicidal run through flames and smoke blanketing Slip No. 1 after the tankship **Markay** exploded. The tanker was beyond saving, so fireboat firemen focused their deck and bulwark rail guns on flames sweeping out from under the American President Lines wharf. From the Collection of Paul Ditzel*

The 10,000 ton *Markay* tankship, typical of the steadily larger petroleum carriers spawned by World War II, was loading butane blend at the Shell Oil Terminal, Berth 166, Wilmington. More than 3 million gallons of petroleum products were on board when, at 2:16 a.m., June 22, 1947, the *Markay* exploded in a Vesuvius of flame and churning clouds of thick, black smoke.

Explosions split the *Markay* completely in half and slashed compartments which vomited hundreds of thousands of gallons of blazing fuel. A blanket of flames fanned across the 600-foot-wide slip and speared underneath the wood wharf supporting two warehouses of the American President Lines. Towering columns of fire lit the port as chunks of the *Markay* rained for miles around. The first alarm was pulled at Firebox No. 15, a mile from the *Markay*. A few hundred yards downstream, the crew of *Fireboat No. 2* was jolted awake by the whooshing explosions which shook their boathouse like an earthquake. They saw the glare and quickly cast off, along with First Mate John Planagan, alone in auxiliary *Fireboat No. 3*.

As Pilot Brainard "Choppy" Gray pulled out of the cavernous boathouse and turned to starboard, fireboat firefighters saw a monumental job dead ahead of them. Only the bow of the *Markay* was visible. The floating blanket of flames had not only involved the underwharf at the Shell terminal, but were touching off the American President Lines' wharf directly across from the *Markay*. Slip No. 1's waters mirrored the flaming eruptions which were gushing still more blazing petroleum products completely across the 600-foot-wide slip.

The fireboat firefighters felt the heat as they approached the slip and saw flames gnawing up and into both American President Lines warehouses which were quickly engulfed. As *Fireboat No. 2* closed upon the fire, its protective spray could not prevent the radiant heat from shattering three pilothouse windows and the starboard lights.

Dozens of land-based fire companies were converging upon the disaster, but there was no question that the fate of the congested area beyond the 1200-foot-long and 600-foot-wide blanket of flames blocking the slip depended upon the crew of *Fireboat No. 2*. Worsening the challenge was the offshore breeze and incoming tide which were pushing the floating flames upstream where a worse catastrophe seemed inevitable. Shrouded by smoke were upstream petroleum storage tank farms, warehouses, a chemical plant, an oil refinery and the huge "20 Mule Team" United States Borax Chemical Corporation complex.

*Fireboat No. 2* would somehow have to plow through that sea of fire and thick smoke to get ahead of the fire, turn around and hit the fire head on if it was going to be checkmated. This was an extremely perilous proposition as all aboard *Fireboat No. 2* realized. The gasoline-powered fireboat was never intended for entering an area with fire on every side of it. The boat's ineffective water spray system proved that. Ignition of gasoline vapors could blow up the fireboat and everyone aboard. Without radar while maneuvering in that thick smoke, the fireboat could slam into a wharf or hunks of the *Markay*. Nor was there effective breathing apparatus on board to protect firefighters from creosote and other toxins in the smoke.

As flames from the *Markay* intensified and the fireboat firemen caught the first whiffs of pungent creosote smoke, Gray signaled the engine room to slow all engines. He held the wheel steady as *Fireboat No. 2* edged closer to the inferno. Heat raised paint blisters along the hull. Gray rang for full speed astern while looking for a spot to plow into the flames and smoke.

Planagan, meanwhile, saw four *Markay* crewmen dive off the stern when the smoke momentarily lifted. Impossible as it was to believe that anyone aboard the *Markay* was still alive, Planagan nosed the small wooden boat, also gasoline-powered, into the withering heat which melted the boat's windshield. He picked up two badly-burned crewmen, but the others disappeared.

Gray and Acting Captain Jack Gordon picked a spot where there was some chance of plowing through the 1200-foot-long patch of fire and smoke to reach open water on the other side. Gray rang for slow ahead. *Fireboat No. 2*'s bow edged into the heat. Overwhelmed by smoke and the stinging creosote smoke which permeated the pilothouse and the engine room, Gray ordered the engines reversed. Suddenly the breeze lifted an edge of the blanket of smoke and the fireboat firefighters seized their opportunity.

*Fireboat No. 2* was immediately swallowed inside the smoke. The crew tried to hold their breaths as the creosote-impregnated smoke burned their eyes, faces and hands. On deck, *Fireboat No. 2* firemen used the bow and other turret guns, as well as the rail-mounted standee nozzles, to sweep flames away from them while the boat bulled ahead. The fireboat suddenly emerged from the smoke and flame. The fire was behind them. Gray quickly brought the boat around and firefighters hit the fire head on with their turret guns and bulwork rail nozzles. The blanket of flames would spread no further. The expanse of industrial and residential districts was saved.

Fires aboard the *Markay* burned for days as did the smoldering ruins of the American President Lines wharf and warehouses. The coroner removed ten bodies from the *Markay*. The two crewmen Planagan tried to save were never found. The heroism and unsurpassed firefighting skills of *Fireboat No. 2*'s crew and Planagan were never recognized with medals. Fire Chief John H. Alderson's reasoning was the same as it was following the World War II *Fredericksburg* affair: "Those fireboat firemen were only doing what they are paid to do." Not until many years later—and a different fire chief—would Medals of Valor go, for the first time, to members of the LAFD fireboat fleet for still another unhesitating act of courage.

From shortly after the end of World War II through 1960 was an unprecedented growth period for fireboat construction in the United States. War's end found much of the nation's fireboat fleet rapidly becoming antiquated. Some cities made-do by purchasing surplus military boats and converting them into fireboats. There were good explanations: Conversions were cheaper and quicker than designing and building new fireboats.

The postwar years, moreover, signified the emergence of the United States as a major world power, militarily and economically. The year 1948 started a lengthy period of high employment, higher wages and increasing volumes of port traffic. Even before that, volumes were being hiked by aid going to war-devastated countries. The Texas City disaster was a case in point. The port had become a major bridgehead for relief aid, including shiploads of ammonium nitrate—such as that aboard the *Grandcamp*—which was used as fertilizer for war-decimated agricultural lands in Europe.

Further adding to port tonnage was Korean War military traffic starting in 1950. When the Saint Lawrence Seaway

"The worst pier fire in the history of New York," was Fire Commissioner Frank Quayle's description of the Grace Lines fire, September 29, 1947. Starting at 8:06 p.m., September 29, 1947, and burning throughout the night and next day, the eight-alarmer destroyed the 823-foot-long pier and the two-story Grace Lines terminal building at the foot of West 15th Street. Among the five FDNY Marine Division fireboats which battled the fire was the 9000 gpm Thomas Willett (lower left). From the Collection of Paul Ditzel

Two years after the Saint Lawrence Seaway gave Great Lakes ports direct access to the Atlantic Ocean, Cleveland beefed up its waterfront protection with the 1961 purchase of the **Anthony J. Celebrezze**. The 6000 gpm at 150 psi capacity fireboat is 60-feet-long, a 16-foot-beam and draws 6-feet. The **Celebrezze** mounts four guns: three near the bow and two at the stern. Carrying 1000 feet of hose, the boat has 16 gated discharge outlets, along with foam firefighting capabilities. From the Collection of Steven Lang

*Philadelphia's 6000 gpm **Bernard Samuel** was the first large fireboat built after World War II. Its water tower, normally nesting in a nearly horizontal position, can be raised 24 feet. From the Collection of Paul Ditzel*

opened in 1959, Great Lakes ports registered more activity with their direct access to the sea and foreign commerce. Altogether, these factors created new and greater port fire protection problems.

Cities large and small were faced with still another consideration. There were only so many war surplus boats available. Most were largely a stop-gap against the recognized need for larger boats especially built for firefighting and the growing demands made upon them to justify their costs: search, rescue, pollution-control and harbor patrol. The problem exacerbated when the U.S. Coast Guard, which had provided primary firefighting and other emergency services along many waterfronts, especially in smaller ports, virtually got out of the firefighting business.

The impact of all of these factors was a 12-year boom in large fireboat construction. From 1948 through 1960, cities bought around 30 boats of more than 1000 gpm capacity; about twice as many as they acquired during the preceding comparable period. Those years also started a trend toward smaller boats; sometimes augmenting the larger capacity fireboats. The explanation was more than the lower cost of small fireboats. Smaller boats were required to cope with the phenomenal growth of recreational boat marinas where the wakes of large fireboats could create damage; if indeed they could enter these shallow waters jam-packed with leisure craft.

Evidence of what was ahead in fireboat development was the National Fire Protection Association's Special Interest Bulletin No. 137, published in 1951. Where large fireboat protection was lacking, or where cities were without any fireboat protection, the NFPA suggested that "the generally preferable type of fireboat is one of 3000-6000 gpm capacity, with twin screws for speed and maneuverability, using die-

sel engines of relatively small horsepower and with one or two monitor nozzles…(with)…a capacity of 4500 gpm at 150 psi, overall length of 58-feet, a beam of 16½-feet and a draft of 4-5 feet."

The first large fireboat built after the war was Philadelphia's *Bernard Samuel,* the model for two other fireboats which would replace the city's antiquated fleet. Contrasting with prevailing fireboat concepts of the time, the *Bernard Samuel* was the predecessor of smaller boats featuring better maneuverability and firefighting capabilities as well as lower than traditionally expected construction costs.

The *Bernard Samuel* was a case of déjà vu. That pioneering Naval Architect William C. Cowles, who was among the nation's foremost fireboat designers, theorized a similar fireboat in 1896. Cowles had postulated that one, two or even three fireboats of relatively small size were better, cheaper and more cost-effective than one large boat. Or, to cite a cliche: Don't put all your eggs in one basket. His colleagues agreed with Cowles, but few fire chiefs listened to him.

The *Bernard Samuel,* despite the fact that Philadelphia had not bought a fireboat for a quarter of a century, was not a rush-to-launch project. The fireboat was an amalgamation of studies of Philadelphia's immediate and long-term requirements, a survey of fireboats in other cities and consultation with the National Board of Fire Underwriters.

Designed by Naval Architect Thomas D. Bowes, the boat was built at the RTC Shipbuilding Corporation, Camden, N.J., across the Delaware River from Philadelphia. The *Bernard Samuel* displaces 94 tons and is 75-feet, 10-inches in overall length, with an 18-foot beam and drawing 4-feet forward and 7½-feet aft. The boat cost under $200,000; about 25 per cent less than that which fireboat builders generally

*More than six alarms called Philadelphia fire apparatus, including the **Bernard Samuel**, to fight flames following an 8:20 a.m., May 26, 1955, series of explosions and resulting fires in the Publicker Alcohol Company. Four workers were killed in the 5-story complex containing vats of alcohol at Delaware Avenue and Bigler Street. From the Collection of Paul Ditzel*

*The **Delaware**, sister fireboat to the **Bernard Samuel** and the **Benjamin Franklin**, all with 6000 gpm capacity, put Philadelphia in the forefront of modern fireboat protection when the boats were delivered between 1948-1950. From the Collection of Bill Dahlquist*

expected to invest at that time.

The *Bernard Samuel* is rated at 6000 gpm at 150 psi. The hull is a modified V-shape, with a large radius at the turn of the bilge. Forward sections are U-shaped. The drag of the keel is about 3½-feet. With a large, balanced rudder, this configuration permits exceptionally fast handling. Design of the well-rounded forefoot section and the bow enables the boat to be driven onto ice or mudflats, along with the ability to back off. This feature would have been impossible if a conventional bar keel and stem with sharp sections forward had been utilized.

The fireboat's welded steel construction is notable for its longitudinal framework, rather than a vertical framing system. This horizontal design provides sturdier protection for the *Bernard Samuel* while operating alongside and around wharves and piers. Additional hull strength includes flat steel bars spaced every two feet on the port and starboard sides, as well as the bottom.

Of particular interest are the *Bernard Samuel's* large fore and aft trim tanks. These permit the reduction of the fireboat's draft while entering shallow waters. Conversely, towing of disabled boats results by using the trim tanks for increasing the draft and, therefore, the boat's heft. Sperry combination electric-and-hand steering was installed for easier, faster and virtually failsafe operation.

Time trials prior to the November, 1948, acceptance of the *Bernard Samuel* registered speeds in excess of 17 miles per hour. At full speed and with the boat's 24-foot mast raised, the *Bernard Samuel* readily turned in a tight circle with only a 5-degree list. The fireboat shifts from full speed ahead to full speed astern in about seven seconds.

The versatility of the *Bernard Samuel* is largely attributed to its powerplant: an 800 horsepower General Motors Series 71 Quad-6. Four 6-cylinder diesel engines are mounted on a common base and drive a single propeller shaft by means of reduction gearing. The shaft turns a 54-inch diameter, four-bladed propeller. This single screw configuration was chosen because it reduces the possibility of damage from floating debris or ice.

The engines can be declutched and switched to pumping mode. One of these engines can simultaneously provide

All three of Philadelphia's fireboats were called to battle a four-alarm fire aboard the freighter, *Polanic*, of Yugoslavian registry. The 6000 gpm *Bernard Samuel* (foreground) was the first fireboat to arrive following the first alarm at 12:04 p.m., June 29, 1961, from Pier No. 3 North on the Delaware River near Delaware Avenue and Market Street. The *Delaware* (left) was sent at 12:29 p.m. The 407-

foot-long freighter, reportedly loaded with rubber pellets, broke loose from its moorings. While fireboats continued to flood the *Polanic*, the freighter drifted across the Delaware River and went aground on a mudflat off Camden, N.J. From the Collection of Paul Ditzel

*The 12,000 gpm* **Deluge** *was built by the Defoe Shipbuilding Company for Milwaukee. The steel-hulled* **Deluge** *was delivered in July, 1949. Diesel-electric powered, the fireboat's overall length is 96-feet, 7-inches and it has a beam of 23-feet and a draft of 6-feet, 9-inches. The* **Deluge** *has four monitor guns, including an aft-mounted gun which elevates 23-feet above the water line. From the Collection of Steven Lang*

propulsion and pumping capabilities. Each GM engine is equipped with a 166-horsepower heavy-duty Rockford clutch for operating the four De Laval two-stage centrifugal pumps which are each rated at 1500 gpm at 150 psi. Utilization of relatively simple machinery and equipment not only reduced the initial investment and later maintenance costs, but permitted operation by a small crew.

As built, the *Bernard Samuel* was equipped with three F.N. McIntyre Brass Works' monitors of 1500 gpm capacity; one of which was mounted atop the Michigan Tool Company elevating 24-foot tower which normally nests in a

nearly horizontal position with the nozzle pointing to the stern. Other firefighting capabilities include six 3½-inch hydrants capable of supplying a dozen 2½-inch hoselines. The *Bernard Samuel*, when commissioned, was outfitted with two Wirt & Knox swivel-type reels carrying 2000 feet of hose. The boat, by 1986, was carrying 1050 gallons of foam concentrate.

Because of the forethought that went into it, the *Bernard Samuel*'s success was a foregone conclusion. It led to the 1950 purchase of two virtually identical fireboats, the *Benjamin Franklin* (better known as the *Franklin*) and the *Delaware*. With these three new boats, Philadelphia stood in the forefront of cities with superior protection. The three boats replaced the 1893 *Edwin S. Stuart*, which was scrapped in 1949 and the 1921 twins, *Rudolph Blankenburg* and the *J. Hampton Moore*. Both were laid up in 1950.

With the *Bernard Samuel*, the floodgates opened for new fireboats in other cities, including the *Victor L. Schlaeger* and *Joseph Medill* (not to be confused with the earlier *Medill* which was taken out of service in 1941) for Chicago; the *Deluge* for Milwaukee—all in 1949—and San Francisco's *Phoenix* in 1954. New York's *John D. McKean* joined the fleet of American fireboats in 1954 and Baltimore's *Mayor Thomas D'Alesandro, Jr.* two years later.

The *Medill* and the *Schlaeger* were the last big fireboats Chicago would purchase; at least through 1988. Virtually identical and of typical Great Lakes low-profile fireboats, they were built for $236,000 apiece by the Christy Corporation, Sturgeon Bay, Wi., following the design of Naval Architect John G. Alden of Boston. Both were powered by eight General Motors 6-cylinder, 2-cycle diesel engines at 200 horsepower developing 2000 rpm. Each boat had six

The **Victor L. Schlaeger,** along with the **Joseph Medill,** were the last big fireboats purchased by Chicago. The 12,000 gpm at 150 psi boats were designed by Naval Architect John G. Alden of Boston, who was best-known for his racing yachts. From the Collection of Paul Ditzel

Chicago fireboats **Joseph Medill** (left) and **Fred A. Busse** attack a 4-11 alarm fire, April 22, 1950, in the Container Corporation warehouse along the Chicago River at 400 East North Water Street, close to the downtown Loop. From the Collection of Paul Ditzel

Chicago's 12,000 gpm **Victor L. Schlaeger** directs its tower and turret gun streams into two 12-story Rock Island Railroad grain elevators leased to the Continental Grain Company. Temperatures hovered at 15 degrees below zero Fahrenheit as a 5-11 alarm and eight special calls summoned apparatus, including 59 engine companies, to this fire at 94th Street and the Calumet River starting at 9:48 p.m., January 21, 1957. From the Collection of Paul Ditzel

Launched February 1, 1954, San Francisco's fireboat was appropriately christened **Phoenix** after the legendary bird which burned to ashes and rose to new life, just as the city did after the Great Earthquake and Fire of 1906. The $392,000 boat delivers 9600 gpm and serves as a floating pumping station to supply the city's auxiliary mains. San Franciscans will never forget that the quake-shattered mains were a major factor in losing a large part of the city to flames. W.C. Dunn

Flagship of the Baltimore Fire Department fleet is the **Mayor Thomas D'Alesandro, Jr.** The 12,000 gpm at 150 psi boat is steel-hulled and dieselized. Designed by Thomas B. Bowes, the **D'Alesandro** was built for $350,000 at the RTC Shipbuilding Corporation, Camden, N.J. The boat is 103-feet, 8-inches long, has a beam of 21-feet, 8½-inches and draws 7-feet, 8-inches. The twin-screw fireboat is rated at 17 miles per hour. Charles Cornell

Chicago's **Joseph Medill** went in service December 2, 1949, and became a familiar sight at its Chicago River berth near the Merchandise Mart. From the Collection of Paul Ditzel

Port and starboard bow turret guns of Chicago's **Joseph Medill** lob streams into a four-story warehouse and light manufacturing plant at 320 North LaSalle Street at the Chicago River. The first call at 2:04 p.m., January 12, 1951, escalated into a 5-11 and three special alarms blaze that brought the response of 25 per cent of Chicago's apparatus, including the **Fred A. Busse,** which joined the **Medill** in fighting the flames. Author of Fireboats was in the rear alley when explosions of illegally-stored flammable liquids caused a partial wall collapse. Three firemen and a member of the Chicago Fire Insurance Patrol were crushed and seven other firefighters were injured. Chicago television stations interrupted programming to make coast-to-coast network feeds. Fire Commissioner Michael J. Corrigan said the $1,500,000 fire was "the biggest I have ever seen in the downtown area." From the Collection of Paul Ditzel

Dean-Hill pumps: twin single-stage centrifugals, each rated at 3000 gpm at 150 psi and four two-stage centrifugals at 1500 gpm. Both the *Medill* and the *Schlaeger* delivered 12,000 gpm at 150 psi.

The all-steel boats were 92-feet-long, 24-feet in beam and had a draft of 7½-feet. Weighing 209 tons apiece, the fireboats had six turret guns, including the gun on the aft-mounted tower which could be elevated 35-feet above the water line. Each boat had 19-gated hose discharge outlets and carried 3600 feet of hose for their frequent use as floating pumping stations to supply land-based engines. Sleeping accommodations were provided for a crew of eight, including an officer, a pilot, two engineers and four firemen. The *Joseph Medill* went in service December 2, 1949, and the *Victor L. Schlaeger* on January 16, 1950.

After battling countless fires for a generation, both fireboats were removed from first-line service in 1986 and berthed at Lake Calumet in South Chicago. The laid up *Schlaeger* will never fight another fire and the *Medill* was on active-standby status. Pragmatically, it is doubtful the *Medill* will answer another alarm from the downtown Loop area or the Chicago River and its industrialized branches. Even when the *Schlaeger* was fully-crewed and in service at its Lake Calumet berth, the boat took one-and-one-half hours to reach the catastrophic January 16, 1967, lakefront fire which destroyed McCormick Place, the nation's largest convention center near downtown Chicago.

Lest anyone think San Francisco will ever forget the Great Earthquake and Fire of 1906, consider the city's decision to acquire a new fireboat to replace the *David Scannell* and the *Dennis T. Sullivan*. Both boats were built after the catastrophe as floating pumping stations in event of another quake-caused loss of water main supplies as well as waterfront protection. Marsh Maslin, a San Francisco Call-Bulletin columnist, asked readers to suggest names for the boat, contracted for in June, 1953. More than 400 were suggested. Thirty readers, including Edward Prendergast of the Phoenix Society, the city's famed fire buff club, came up winners. All suggested the name, Phoenix, the legendary bird which, like San Francisco, burned to ashes and rose to live again.

No better name than *Phoenix* could have been chosen for San Franciso's $392,000 fireboat designed by John G. Alden, Inc., the Boston naval architectural firm which was world-famous for its racing yachts. Alden, who designed several fireboats of the period, came up with an all-welded steel, diesel-powered boat which measured just over 88-feet in length, with a beam of 19-feet, 4-inches and drew 5-feet, 11-inches. The boat, weighing only 80 tons without equipment, was built at the Plant Shipyards Corporation, Alameda, and launched at 11 a.m., February 1, 1954.

Alden provided propulsion and pumping by choosing Cummins Engine Company diesels. Two Model NVHMS V-12, 550 horsepower at 2100 rpm main engines drive propellers or pumps. A third Cummins of the same type is dedicated to pumping. The main propulsion engines each drive the two shafts which are connected to 40-inch-diameter, 24-pitch, Columbian Bronze Company propellers. Link Belt Company chain drives connect the engines to the propeller shafts.

The **Phoenix** of San Francisco has a 9600 gpm at 150 psi capability and mounts four guns, plus hose discharge outlets forward of the tower. The 50-foot telescoping tower gun is rated at 3000 gpm. From the Collection of Bill Dahlquist

Los Angeles' **Fireboat No. 2** cools endangered tanks during a three-alarm fire, June 29, 1954, at the San Pedro Marine Terminal of Tide Water Associated Oil Company. Starting at 6:25 p.m., explosions and fires quickly engulfed 11 of the 17 storage tanks in the facility. Moored beside the boat is **Fireboat No. 3.** From the Collection of Bill Dahlquist

The *Phoenix*'s twin rudders are instantly responsive to the Vickers, Inc., hydraulic steering gear. A Westinghouse Air Brake Company pneumatic system puts control primarily in the pilothouse, with secondary operation in the engine room. When built, the boat could reach a speed of 15 miles per hour. Two auxiliary Cummins 110 horsepower Model HMR-400 diesels each drive an Imperial Company generator for supplying the boat's electrical needs. These auxiliaries, moreover, supplement the main engines when they are switched to pumping mode by holding the *Phoenix* on station while its monitors are operating.

The *Phoenix*'s three De Laval Turbine Pacific Company pumps are two-stage centrifugals, each rated at 3200 gpm at 150 psi. Salt water intake is via two sea chests which supply the 10-inch-diameter suction line. The pumps feed to a central loop leading to four F.N. McIntyre Brass Works' monitors of between 2000-3000 gpm capacity, including the 3000-gpm gun mounted on a Rotary Lift Company 50-foot telescoping tower near the stern. The hydraulic jack-operated tower can be elevated at the rate of 29-feet per minute.

Additional firefighting capability was provided by 14 hose outlets forward of the tower. The boat carried 3800 feet of hose in fore and aft bins. In recognition of the petrochemical traffic in the port area, Alden outfitted the *Phoenix* with a 500 gallon National Foam System, Inc., outfit with four 3-inch nozzle outlets as well as monitor gun dispersal capability. As installed, the system could discharge 165,000 cubic feet of fire extinguishing agent.

When new and fully-equipped, the *Phoenix* weighed 126 tons and had a crew nine: an officer, a pilot, two marine engineers and five firefighters. Cognizant of the historic quake which severed water main systems, the *Phoenix* can pump San Francisco Bay water into the city's auxiliary mains feeding the Twin Peaks emergency reservoir four miles from the waterfront.

Shortly after the launch of the *Phoenix*, New York also had a new fireboat, the *John D. McKean*, the city's first since the 1938 *Fire Fighter*. At $1,426,000 the *McKean* was, at that time, the costliest fireboat ever built. Duplicating it in 1989, however, would likely cost at least three times more. It was big: more than 334 net tons, with an overall length of 129-feet, 9-inches, a 30-foot beam and a maximum draft of 9½-feet. The 19,000 gpm *McKean*, commissioned May 1, 1955, marked the end of extra-large capacity fireboats, at least through 1989.

The *McKean*, packed with innovations, was designed by John H. Wells, Inc., of New York and was built by the John H. Mathis Company at its Camden, N.J., yards. The welded steel boat has twin propellers recessed into modified tunnels. These enable the 12-mile-per-hour *McKean* to enter water without risking damage to the propellers from flotsam and jetsam. The fireboat's extraordinarily large three-bladed propellers—5-feet, 3-inches in diameter—are of solid manganese and made by Ferguson Propeller and Reconditioning Company. These extra-large propellers also increase the *McKean*'s efficiency.

Push button control from the pilothouse enables the *McKean* to instantly double the speed of its rudder response. As innovative as push button control was at that time, there was a major payback in fast maneuverability when emergency situations arose. These included on-station firefight-

The Fire Department of New York's 19,000 gpm **John D. McKean** is not only one of the largest fireboats ever built, but probably marked the end of the era of jumbo fireboats. Commissioned May 1, 1955, the steel-hulled, diesel-electric powered boat was named in memory of FDNY Marine Engineer John D. McKean who was fatally scalded, September 22, 1953, in an engine room explosion aboard the half-century-old **George B. McClellan**. It is believed that the boiler explosion occurred during a water display demonstration. Jack Lerch

**Smoke II**, commissioned April 29, 1958, primarily served as a launch and protection for small boat harbors by the Fire Department of New York's Marine Division. The steel-hulled boat built by Equitable Equipment Company, New Orleans, was similar to those which serviced oil drilling rigs in the Gulf of Mexico. As built, the diesel-operated fireboat had a 600 gpm at 150 psi pumping capacity. Before the boat was sold, December 16, 1974, its pumping capacity was increased to 2000 gpm. **Smoke II** was 51-feet, 9-inches long, had a beam of 14-feet and a draft of 3-feet, 10-inches. It displaced 34½ tons. From the Collection of Paul Ditzel

*New York's **John D. McKean** battles a two-alarm fire, September 2, 1969, at Brooklyn's 27th Street Pier. By 1989, the 19,000 gpm **McKean** continued to serve as the flagship of the FDNY Marine Division fleet. FDNY*

ing or, perhaps recalling the *El Estero* fire during World War II, quick, sensitive navigational response which is mandatory when towing burning vessels from potentially catastrophic situations. The fail-safe system permits manual steering in event of breakdown.

Still another break with tradition was the separation of propulsion and pumping capabilities. The old tried-and-true system was for the engines to perform these dual functions. The result was a sometimes Rob-Peter-To-Pay-Paul situation where an awkwardly delicate and sometimes dangerous tradeoff had to be struck between propulsion and pumping.

The new boat's name evidenced how far fireboat technology had advanced in less than a century. The diesel-electric *McKean* was named in memory of FDNY Marine Engineer John D. McKean. He was fatally scalded, September 17, 1953, by an engine room explosion which wrecked the 7000 gpm antique, *George B. McClellan*, a half-century-old coal burner.

The *McKean*'s twin propulsion engines are direct-revers-

ing, direct-connected Enterprise Engine and Machinery Company Model MG-38, 4-cycle, supercharged diesels; each rated at 1000 horsepower at 425 rpm. The two diesel pumping engines, also built by Enterprise, are Model DSG-316, 4-cycle turbocharged, 6-cylinder units rated at 1000 horsepower at 800 rpm. Electric generating systems were by the Buda Company and Delco Products. Diesel air compressors were supplied by Buda and the Worthington Corporation. The four 4500 gpm at 150 psi pumps were also supplied by Worthington. Tests show the pumps can deliver over 19,000 gpm.

Three sea chests located on the port and starboard hull sides as well as at the bottom of the engine room's forward end, are cross-connected by a 27-inch-diameter main with valves. The main feeds 14-inch and 21-inch mains to each pair of pumps. The pumps also supply six 3000 gpm monitor guns by M. Greenburg & Sons. One of these guns is mounted on a 39-foot tower to give it a 47½-foot height above the water line. Sixteen hose outlets are evenly distributed and manifolded fore and aft of the main deckhouse. Four hose reels can carry 4500 feet of hose.

On October 14, 1960, Baltimore put in service three new fireboats: the **P.W. Wilkinson**, the **Mayor J. Harold Grady** and the **August Emrich**. The 6000 gpm at 150 psi fireboats cost $350,000 apiece and were designed by Thomas B. Bowes, with construction at Jackobson Shipyard, New York. They are 85-feet-long, have a beam of 20-feet, draw 8-feet and are rated at 15 miles per hour. From the Collection of Bill Dahlquist

New Haven's **Sally Lee** is one of the few fireboats named after a woman. Sally Lee, daughter of Mayor Richard C. Lee, was honored when the boat was christened in 1962. The boat, designed by John G. Alden, Inc., of Boston, was built at the Norfolk Shipbuilding & Drydock Company, Norfolk, Va. The welded steel boat is diesel-operated and includes two single-stage De Laval centrifugal pumps, each rated at 3500 gpm at 150 psi. The **Sally Lee** is 68¹/₂-feet-long, had a 20¹/₂-feet beam, draws 5-feet and displaces 70 tons. The fireboat is equipped with four guns and carries 760 gallons of foam concentrate: sufficient to provide 129,000 square feet of fire extinguishing agent in 23 minutes, says Assistant Chief John E. Smith of the New Haven Fire Department. John G. Alden, Inc.

Many honors were heaped upon Astronaut John H. Glenn, Jr., the first American to orbit Earth on February 20, 1962. Among them was the Fire Department of New York's fireboat **John J. Glenn, Jr.**, commissioned September, 13, 1962. The 5000 gpm at 150 psi boat was designed by H. Newton Whitteley, Inc., New York and built at the Diesel Shipbuilding Company, Atlantic Beach, Fla., at a cost of $230,000. The boat's specifications include its 70-foot length, beam of 21-feet, draft of 5-feet and 83 ton displacement. The **Glenn**, which had four deck pipes, was sold to Washington, D.C., in 1977. The District of Columbia Fire Department purchased the boat for $150,000 and increased its capacity to 7000 gpm at 150 psi. At welcoming ceremonies, Senator Glenn (Dem. Ohio) was named Honorary Fire Chief of the DCFD in recognition of his space explorations and his congressional leadership in fighting the nation's arson problem. The **Glenn** is the only fireboat named in honor of an astronaut. From the Collection of Paul Ditzel

John G. Alden of Boston designed the **City of Portland,** a 7000 gpm at 150 psi fireboat for Portland, Me. Built in 1959 for $238,481, the welded steel boat is 64-feet, 10¹/₂-inches long, had a 18¹/₂-foot beam and draws 5-feet, 11-inches. The **City of Portland** was to see frequent use in emergency medical service to five Casco Bay islands which are part of Portland. The boat takes medics to the islands and returns with their patients. John G. Alden, Inc.

If it is true that every cloud has a silver lining, consider the massive clouds of ugly black, toxic and severely injurious creosote-impregnated smoke that chuffed from under the Matson Lines wooden wharf, St. Patrick's Day, 1960, in Los Angeles. The fire had spread to a lumber barge when the first alarm was received at 3:42 p.m., March 17. Also endangered was none other than the *Hawaiian Rancher,* the same Matson freighter which burned in San Francisco Bay, December 1, 1948, and was the first fire fought by Oakland's fireboat, *Port of Oakland.* The *Rancher* quickly put to sea.

Flames spread from Berth 200A, which was under construction, to the underside of Berth 199. *Fireboat No. 2* opened up its Big Bertha, deck and other guns. Despite tens of thousands of gallons the fireboat shot into the fire,

*Guns of Los Angeles **Fireboat No. 2** pound flames which destroyed 1100 feet of wooden wharfing during this St. Patrick's Day, 1960, fire at the Matson terminal. Wooden gridwork under the wharf made it impossible for streams to effectively penetrate the fire. The Matson blaze led to the development of new firefighting tactics which utilized SCUBA divers who take hoselines directly under burning wharves. George E. Bass*

The first fireboat equipped with an articulated, elevating platform is Toronto's **William Lyon Mackenzie**, a 7000 gpm at 150 psi boat launched November 7, 1963. Starting at 2:22 p.m., July 21, 1965, the **Mackenzie** fought its first major shipboard fire when an alarm was sounded from Pier 11 for a blaze aboard the 7650-ton **Orient Trader** loaded with crude and latex rubber. The **Mackenzie**, berthed only 600 feet from the Greek freighter, had water on the fire within minutes. As the blaze threatened nearby warehouses, the **Orient Trader** was towed into Humber Bay, near Ward's Island, and grounded. Firefighting continued as the **Mackenzie's** platform-mounted 4-inch nozzle delivered 2000 gpm. Foam and water controlled the fire by 7 a.m., the following morning. From the Collection of Paul Ditzel

Los Angeles Fire Department SCUBA divers train for underwharf firefighting as they leap overboard from **Fireboat No. 3** with their pontoons and hoselines. From the Collection of Paul Ditzel

Forming a water curtain with their nozzle-mounted pontoons, teams of Los Angeles Fire Department SCUBA divers use the back pressure in hoselines to propel them under wharves. From the Collection of Paul Ditzel

Marine Architect William Francis Gibbs, who designed the famous fireboat, **Fire Fighter**, for the Fire Department of New York, also designed the only land-based firefighting system with fireboat capabilities. Gibbs' concept became reality when, on October 1, 1965, the FDNY put the Super Pumper System in service. Nub of the system was the Super Pumper, built by Mack, which delivered 8800 gpm at 350 psi. Suctioning from waterfront supplies or inland hydrants, the Super Pumper could supply 35 hoselines. In addition to the Super Pumper, the system included a Super Tender, plus three Satellite hose and water cannon rigs. This most powerful fire engine ever built did not enjoy the longevity of the **Fire Firefighter** and was retired October 25, 1982. From the Collection of Paul Ditzel

Fog stream from nozzle mounted on a pontoon enables Los Angeles SCUBA fireman to direct stream straight up to hit pockets of fire which could not otherwise be reached with fireboat streams or those from land-based engines. From the Collection of Paul Ditzel

the pilings and timber latticework of 10-by-20-inch wood supporting the thick deck barred the streams from penetrating to the heart of the flames. *Fireboat No. 2* held close to the wharf despite the thick smoke which sometimes enveloped the boat and forced firemen in the engine room to put on breathing apparatus.

The situation was no better on the land side where firemen from 21 engine and five ladder truck companies were trying to stop the flames. Imminently threatened was a 128-foot-tall gantry crane at the $10 million Matson passenger terminal. Firemen fought to save the crane by concentrating cooling streams on its legs. Dozens of firefighters took the worst smoke beating of their careers as they battled to checkmate the fire under the 4-inch thick, asphalt-covered deckwork.

Firemen used jackhammers, iron bars and chain saws to cut holes in the deck so Bresnan and similar distributor-type nozzles could spin circular patterns of water under the wharf. These streams were as moderately effective as those of *Fireboat No. 2*. The lumber barge burned loose from its moorings. Adrift, a towline was secured to it and the barge was taken to shore where firemen saved the barge, but not the lumber.

The Saint Patrick's Day fire burned on into the night and was not controlled until the next day after destroying 1100 feet of wharf. The $350,000 crane and the Matson terminal was saved. Once again the historic nemesis of waterfront firefighters: underwharf and dock fires, showed the American fire service had made slight progress in quickly and effectively controlling them. Tactics had changed little since the first modern fireboats were built in the late 1800s. Neither fireboat guns nor cutting holes in deckwork produced quick results. Streams operated from portholes in fireboat hulls were only a bit more effective.

Some fireboats, notably San Francisco's *Phoenix*, carried a punt. Firemen lowered these small boats into the water and maneuvered close to underwharf and dock fires while using hoselines supplied by the fireboat. But, as always, effective fire control was stymied by thick spiderwebs of wooden underpinning which concealed hidden pockets of fire spreading along and under wharves and docks.

The Matson fire kindled the imagination of Assistant Chief William W. Johnston, Jr., of the LAFD, who came up with an idea which was to revolutionize underwharf and dock firefighting tactics around the world. Johnston admitted he was no waterfront firefighting expert, but for 15 years he had been an expert skin-diver. He suggested that recreational skin diving techniques could be adapted to firefighting. Wet-suited SCUBA (self-contained underwater breathing apparatus) equipped firemen could, like frogmen, flipper across the water while taking hoselines under burning piers and wharves.

Using adjustable straight-stream or fog-like spray nozzles, Johnston theorized that skin-divers could do what fireboats and firefighters on docks could not: quickly and effectively block flames and hit those infamous and inaccessible pockets of fire. Johnston believed SCUBA firemen could have prevented most of the $65 million loss at the Matson fire. "Within the first 30 minutes, I'm positive scubamen would have confined that fire," he said.

LAFD officials publicly said they encouraged ideas, however bizarre, that would improve waterfront firefighting tactics. Privately, they thought it was a crackpot idea, but nevertheless provided limited funding for early experiments. Except for Johnston, most chiefs were understandably skeptical about the notion of sending firemen under burning wharves where falling debris could injure or kill them. Melting tar and creosoted smoke could, moreover, cause severe burns or asphixiation. How could fireboat firefighters avoid hitting frogmen firefighters, hidden in smoke, with a lethal monitor gun stream? How many firemen would voluntarily jump into six fathoms of water and flipper into areas of tremendous smoke and flame? One hundred twenty LAFD firemen, mostly skin diving enthusiasts, applied when Johnston put out a call for volunteers to develop SCUBA firefighting equipment and techniques.

The program proceeded slowly, much of it during the firemen's off-duty time. However much skin diving experience they had, all candidates had to demonstrate their ability to swim long distances, carry weights on their backs and float a specified length of time. Training of certified LAFD SCUBA instructors included that of the National Association of Underwater Instructors.

Cold water suits helped to prevent hypothermia while divers were in the water and also prevented cuts and abrasions while the firefighting skin-divers unavoidably rubbed against pilings and cross-braces. The U.S. Divers Company provided small, lightweight air tanks equipped with a mouthpiece. Skin lubricants, applied to exposed parts of the divers' faces and hands, helped to protect against creosote burns.

Training progressed from swimming pools to offshore ocean waters and finally into Los Angeles Harbor. Johnston and his divers were surprised to find that 2½-inch diameter hoseline floated as easily as 1½-inch diameter lines. Equally surprising was another discovery: When fireboats pumped into the hoselines, the weight of the water did not sink the hose which could, moreover, be extended for greater lengths than expected. The explanation was that the weight of water in the hose was neutralized when the lines were in the water.

Concurrent experiments developed suitable nozzle-mounted pontoons: Inner tubes, hollow sheet metal floats and balsa wood rafts proved unsuitable. Johnston and his divers evolved a two-by-three-foot float of dense polyurethane foam covered with two layers of bright yellow fiberglass for optimum operating conditions, better visibility and safety. Hand grips were recessed into the sides of the floats.

Each float weighs 28 pounds, including the 11-pound center-mounted nozzle that points straight up. Each of these nozzles delivers 300-500 gpm at 150 psi. Linked by hoselines, chains of floats provide single-attack capability as well as water curtains for establishing fire barriers under docks. A smaller float, to which 1½-inch diameter hose is attached, is used for both attack and mopup of lingering pockets of fire. Their nozzles deliver around 100 gpm at 50 psi.

SCUBA firefighting resulted in new underwharf and dock fire tactics. When an alarm indicates an underwharf or dock fire, skin-divers aboard fireboats suit up enroute. Upon arrival they take to the water with their floats and

*Boston's 6000 gpm **Joseph J. Luna** battles a four-alarm fire which began at 1:01 p.m., November 30, 1961, at Pier 17, Castle Island terminal. The stubborn fire under the pier burned for four days before it was declared out. Boston SCUBA firemen not only helped to fight the blaze, but assisted with mopup after the fire was controlled. From the Collection of Bill Noonan*

*Flames and creosote-permeated smoke roll skyward from under the blazing substructure of Pier No. 7 at Sitcum Waterway and 11th Street in the Port of Tacoma. The 3-11 alarm fire, starting at 2:34 p.m., Sunday, July 14, 1963, was wind-driven faster than a person could walk. The rapid spread overwhelmed firefighters on the pier. Battalion Chief Arthur Strong was killed while making certain all his men were safe. The 10,000 gpm **Fireboat No. 1** joins the firefight which marked the first time SCUBA firefighting techniques were used in Tacoma. TFD*

*Los Angeles purchased **Fireboat No. 4** as part of a master plan for long-range port protection. The $639,000 boat is rated at 9000 gpm at 150 psi and carries 550 gallons of foam solution for petrochemical fires. Its superb maneuverability features includes the capability of moving side-to-side and back-and-forth by means of jet-stream nozzles. The boat, commissioned in 1962, was later renamed the **Bethel F. Gifford** in honor of the late LAFD battalion chief whose research and efforts were primarily responsible for the fireboat's innovations. From the Collection of Bill Dahlquist*

*In 1966, Los Angeles contracted for three 34-foot-long high-speed fireboats. They have the capability of operation from the cabin as well as a flying bridge. The boats were designed for small boat marina protection, as launching platforms for SCUBA divers and as waterfront fire command posts. From the Collection of Jim Perry*

hose, swim clear of the boats and then open their nozzles. Divers propel themselves by utilizing the backward thrust of the nozzle pressure. To move ahead, they depress the back of the float, thus changing the angle of water discharge from vertical to over their shoulders. To turn right or left, divers tip the sides of the float. Turning right merely requires depressing the left side of the float. Conversely, leftward travel results from dipping the float's right side. Divers rarely have to submerge and the water pressure virtually eliminates the need for flippering while traveling across the water.

SCUBA firefighting or rescue work is closely coordinated among the two-man teams which use the traditional skin diving buddy system, while backup divers on the fireboats provide additional safety. LAFD SCUBA techniques usually starts with the divers propelling themselves under the dock on each flank of the fire. They may set up water curtains to help prevent the spread of flames. Fireboats, meanwhile,

bore into the areas between the flanks. When the fire is knocked down, the fireboat guns are shut off and the SCUBA divers propel themselves in from the right and left flanks to put out pockets of fire inaccessible to fireboat or land-based streams.

Johnston and his enthusiastic volunteers largely won over chief officers, but many of them remained doubtful. Theory and drills were one thing. How effective would SCUBA firefighting prove when a major underwharf fire occurred? That question would not be answered for several years as Johnston and his skin divers continued to refine their techniques.

Other cities, meanwhile, were developing SCUBA firefighting programs. Fourteen Boston Fire Department SCUBA divers, using only eight 1½-inch diameter hoselines, prevented the November 30, 1961, spread of a fire under the Pier 17, Castle Island, terminal. When Tacoma's Pier 7 burned, July 14, 1963, the department called for Fireman Bob Beatteay of the nearby Renton, Wa., Fire Department. Beatteay, an experienced skin-diver who exchanged ideas with LAFD Chief Johnston, had developed underwharf firefighting tactics. Tacoma officials were so impressed by these tactics that they started a SCUBA firefighting program.

Five years after Johnston and his fellow LAFD SCUBA firefighters began developing their techniques, no major underwharf fire had occurred. With official interest flagging and continued support doubtful, Johnston contacted a friend, Paul Ditzel, a national magazine writer and contributing editor of Fire Engineering, a world preeminent fire service publication.

Ditzel's front-cover story in the November, 1965, issue of Fire Engineering was headlined, "Los Angeles Fields SCUBA team", and was highlighted by a full-page photo of LAFD skin-divers leaping into the water. The article enthusiastically supported Johnston's ideas. The assistant chief was deluged with inquiries from fire departments around the world. With such worldwide acclaim for the LAFD's program, funding could hardly be stricken from the LAFD budget.

As SCUBA training continued toward the inevitable fire which would give LAFD divers the chance to prove their value, the department proceeded with a long-range master plan for protecting the Port of Los Angeles. Traditional freight handling methods were being supplanted by cargo container vessels and related facilities. Hazardous materials installations seemingly were proliferating as much as the materials themselves. Small boat marinas were growing and had their distinct protection needs, including the ability of fireboats to enter these waters without creating dangerous wakes which could damage leisure craft.

By 1987, the phenomenal growth of traffic volumes and property values in the Port of Los Angeles would be evidenced by the more than 3500 vessels, including cargoships, tankers, passenger liners and lumber carriers, that arrived in the port. In that year Los Angeles would register 17 million tons of general cargo and 20 million tons of bulk oil products being handled along the 28-mile waterfront. While concrete piers and wharves would be built for new installations, nearly half of the port's 13 miles of wharves would remain wooden by the end of 1989.

The LAFD's plan to meet these and other port protection requirements included modernization of Fireboat No. 2, the purchase of three small fire-rescue speedboats and the $639,000 investment in a new fireboat of 9000 gpm capacity: Fireboat No. 4. That boat was to become a benchmark in the development of American fireboat technology.

First of all, it is relevant to note that the history of American fire apparatus manufacturing is replete with the names of builders whose firefighting experience was incorporated into the design and construction of fire engines and ladder trucks. Fireboat design and construction, on the other hand, traditionally falls within the expertise of naval and marine architects and designers without firefighting experience. This is, of course, attributable to the sophisticated art and craft of shipbuilding design and construction. The market for large fireboats is, moreover, small. Relatively few American cities require them. Less than 200 fireboats of 1000 gpm or more capacity have been built during the 20th Century.

Fireboat No. 4 resulted from a blend of the waterfront fire protection experience of Battalion Chief Bethel F. Gifford, whose research melded with the design team of L.C Norgaard & Associates, San Francisco, and the builder, Albina Engine & Machinery Works, Portland, Ore. Gifford's interest up to that time was paralleled only by that of New York Fire Chief Edward F. Croker around 1900. It was not to be repeated until the early 1970s when Chief Tony F. Mitchell of the Tacoma Fire Department took a profound interest in fireboats which resulted in the first Surface Effect System multi-purpose fireboats which travel on a trapped bubble of air.

Los Angeles' Fireboat No. 4 has a welded steel hull and is powered by six Cummins diesel engines. Two Model VT-12M engines are dedicated to propulsion. Each develops 600 horsepower at 2100 rpm and is shafted to controllable pitch propellers. Two of the same model Cummins are coupled to a pair of a De Laval two-stage centrifugal pumps rated at 3000 gpm at 150 psi. In addition to that 6000 gpm capability, two Cummins NRT0-6-M engines of 335 horsepower at 2100 rpm are each linked to auxiliary De Laval two-stage centrifugal pumps of 1500 gpm capacity apiece. Two 50 kilowatt General Electric generators power the boat.

Fireboat No. 4 is equipped for conventional as well as petrochemical firefighting. Its most powerful gun is atop the pilothouse and delivers 3000 gpm at 150 psi. Bow and stern guns are each rated at 2000 gpm at 150 psi. Two port and starboard monitors put out 1500 gpm at 150 psi apiece. Additionally, twelve bulwark rail-mounted nozzles on the port and starboard sides each deliver 500 gpm at 150 psi in variable straight or fog spray streams.

For underwharf and dock firefighting, Fireboat No. 4 has two 2500 gpm at 150 psi nozzles just above the waterline and located abaft the port and starboard bow. These Stang articulating arm nozzles are hydraulically-operated from the pilothouse. The boat can supply hoselines to land-based firemen via 15 gated discharge outlets. It carries 1850 feet of hose in two aft-mounted boxes. A major capability of Fireboat No. 4 is its 550 gallon tank of foam solution for petrochemical fires. Depending upon needs, the boat can provide up to around 40 minutes of foam dispersing capabilities.

Maneuverability is prominent among the boat's features. The controllable pitch propellers not only instantly respond to forward and reverse movement, but can help turn the boat in a complete circle inside its 76½-foot length. Fireboat No. 4 can, moreover, maneuver in a square-box-like fashion: side-to-side and back-and-forth. This is accomplished by four jet-stream 1000 gpm nozzles at 80 psi. They are evenly distributed at the port and starboard bow and stern. By direct pilothouse control and the variable pitch propellers, the jets similarly hold the boat on station, thus solving the traditional problem of nozzle back pressure pushing the boat away from its target.

Fireboat No. 4, with Gifford aboard, arrived in Los Angeles from Portland, February 22, 1962, with a welcoming escort of fireboats and other harbor craft. It was one of the most spectacular civic celebrations in Port of Los Angeles history. The boat has a crew of five: a captain, a pilot, two engineers and a firefighter. Appropriately, the boat was renamed the Bethel F. Gifford on California Fire Service Day, May 8, 1965, in memory of the chief who had played so vital a role in her innovative design.

*Los Angeles Fire Department SCUBA divers got their first chance to demonstrate their capabilities when an underwharf fire involved the Harbor Grain Terminal at Berth 174 on December 28, 1967. More than 24 SCUBA divers made around 90 sorties under the wharf and helped to keep the loss to only 370 feet of the 3700-foot-long wharf. From the Collection of Paul Ditzel*

Completing the new look in the LAFD fireboat fleet were three 34-foot-long, high-speed planing hull-type boats built by Drake Craft, Inc., Oxnard, CA. The boats were designed for fires and other emergencies in small boat marinas, as launching platforms for SCUBA divers and as waterborne command posts.

The December 3, 1966, contract for the boats, costing around $62,000 apiece, called for a fiberglass-over-wood V-bottom hull design to enable them to develop fast, but safe speeds in marinas. As fast initial attackers, the boats have a top speed of 30 miles per hour, but slower speeds are mandatory in small craft harbors. Each boat has a flying bridge atop the cabin to permit flexible dual operational control areas.

As built, the boats were gasoline-powered by three water-cooled Chrysler Corporation Model M-426-B Imperial 290 Marine engines developing 290 horsepower at 4000 rpm. The two outboard engines provide propulsion. The center engine drives a single-stage Waterous Model CGNF pump rated at 750 gpm at 150 psi. Each boat mounts a bow gun with a 175-foot-reach. The craft carry 65 gallons of foam concentrate which can result in the production of around 17,000 gallons of foam.

With its new and modernized fireboat fleet, the LAFD was confidently prepared for current and projected protection needs of the port as supertankers, some of them more than 1100-feet-long and carrying around 2 million barrels of petroleum products, began making Los Angeles a frequent port-of-call. With all the attention given to the modernistic port, LAFD's SCUBA firefighters began to lose hope that they would ever be called upon to fight a major fire under a wooden wharf and justify their seven years of preparation.

They were the first fully-equipped fire department SCUBA divers in the world and probably the last to get a chance to show their mettle.

When they were finally called to action it was not for a moderately-sized fire, as they expected, but a full-blown underwharf blaze, December 28, 1967, which also involved a large transit shed at the Berth 174 Harbor Grain Terminal. Quickly following the first alarm at 10:32 a.m., additional calls brought more fireboats, 45 engine companies and five ladder trucks as the fire worsened.

Battalion Chief Jack S. Douglass took command of marine operations aboard *Fireboat No. 5*, while *Fireboat No. 2* and *Fireboat No. 4* simultaneously bored into the underwharf and transit shed smoke and flames with their monitor and underwharf guns. SCUBA divers lept overboard and attacked from the flanks. Their floats' nozzle pressures propelled them under the wharf. While holes were being punched in the deck to reach the fire from above, off-duty SCUBA divers were called. More than two dozen divers proved the value of SCUBA firefighting that day as they made around 90 sorties under the wharf. Dangerous as their work was, only one of them, SCUBA Diver Charles C. Hilzer, was injured and he would soon return to duty.

The Harbor Grain Terminal, for all its disaster potential, was soon controlled by the teamwork of water and land firefighters. Losses amounted to a relatively small $150,000. Of more interest to Johnston and his SCUBA divers was the fact that only 370 feet of the 3700-foot-long wharf burned before the fire was stopped. Never again would anybody consider SCUBA firefighting a harebrained idea that would never prove viable.

*At 10:32 a.m., December 28, 1967, an alarm from the Berth 174 Harbor Grain Terminal called Los Angeles'* **Fireboat No. 4** *(foreground) and* **Fireboat No. 2.** *Fireboat and land-based firemen were confronted upon their arrival with a full-blown wooden underwharf fire which had extended into a large transit shed. From the Collection of Paul Ditzel*

# CHAPTER EIGHT
## GALLANT SHIP

*Wind and tide spread flames as long as 30 football fields and higher than a 10-story building after the Sea Witch slammed into the Esso Brussels tanker near the Verrazano-Narrows Bridge in New York, June 2, 1973. Frank Duffy*

*Flames which engulfed the Sea Witch (top) and the Esso Brussels tanker were controlled by dawn, June 2, 1973, in the Port of New York. The FDNY Fireboat John D. McKean (far right) is joined by a Moran, U.S. Coast Guard and Exxon tugs in cooling the vessels. Frank Duffy*

Ship traffic in the Port of New York was proceeding routinely shortly after midnight that Saturday, June 2, 1973. Visibility was excellent, a westerly wind was blowing and the current was strongly at ebbtide. The American Export Isbrandtsen ship, *Sea Witch*, was putting to sea with 730 loaded cargo containers on board; nearly 300 of them stacked on deck.

The *Sea Witch*, considered a jinx ship because of her frequent navigational and other breakdowns, was doing a fast 13 knots as she approached the Verrazano-Narrows Bridge connecting Staten Island and Brooklyn. Suddenly the *Sea Witch* lost steering control and veered to starboard. Dead ahead was the Belgian tankship, *Esso Brussels*, at anchor north of the bridge and parallel to the Staten Island shoreline. Aboard the 14,500 - ton tanker were 2 million barrels of highly-volatile Nigerian crude oil.

The master of the *Sea Witch*, John L. Paterson, and his crew desperately tried to regain control as the 581-foot-long containership's bow pointed toward the starboard stern of the 655-foot-long *Esso Brussels*. Paterson ordered the *Sea Witch's* engines full astern and the port anchor dropped to help slow the out-of-control ship. The port anchor would not go down. Neither would the starboard anchor. Realizing that a collision was inevitable, Paterson ordered the *Sea Witch's* whistle to start sharp, short, warning hoots. As the *Sea Witch* furrowed closer to the *Esso Brussels*, the warning hoots became a continuous shriek. The shipboard alarm was sounded and a Mayday call put out.

The *Sea Witch*, its whistle shrieking, slammed into and impaled the *Esso Brussels* at Nos. 7 and 8 starboard tanks about midway between the forward and afterhouse. The resounding impact tore away the *Sea Witch's* starboard anchor and flung it onto the deck of the tanker. The *Sea Witch's* bow cut a 15-foot wedge into the *Esso Brussels*. More than 31,000 barrels of oil gushing from the slashed compartments instantly ignited.

The wind and the strong ebbtide spread sheets of flame as long as 30 football fields and as high as a 10-story building. The still spinning propellers of the *Sea Witch* churned the flames out into the channel as both ships drifted under the Verrazano-Narrows Bridge. Radiant heat scorched its underside while clouds of smoke stopped all traffic over the 4260-foot-long bridge, the longest in North America.

Paterson, one of the first to perish, fell dead of a heart attack. Two of his crewmen died in the flames. Dozens of others fled to the after deckhouse. Similar pandemonium was aboard the *Esso Brussels*. Unable to lower lifeboats as flames quickly encircled the tanker, many crewmen grabbed life preservers and jumped overboard. Their reflective orange preservers and their upper bodies were instantly smudged black by oil. The current carried many of them far into the channel where at least 10 drowned or died from burns.

Calls deluged FDNY alarm offices where dispatchers, recognized as among the world's best, wisely marshaled firefighting and disaster resources on both sides of The Narrows by transmitting a second alarm on the Staten Island side and a third alarm at the Brooklyn waterfront. The nearest fireboat, *Fire Fighter*, answered the first alarm at 12:46 a.m., which reported a burning ship near the Verrazano-Narrows Bridge. Leaving their Staten Island berth, the *Fire Fighter's* 10 crewmen could not possibly know that not one but two ships were buried in the vortex of towering columns of smoke and fireballs erupting from The Narrows.

As they approached the port side of the blazing *Esso Brussels*, the *Fire Fighter*'s crew used their deck pipes to sweep aside the floating flames while using the boat's searchlights to look for survivors and make certain none of them were snagged by the *Fire Fighter*'s propellers. Knocking down the *Brussels'* port side flames, the *Fire Fighter* came around to initiate a starboard attack. For the first time they saw that a second vessel was burning.

The *Fire Fighter*'s deck pipes bored into fires aboard both ships as Pilot Matthew T. Fitzsimmons, Jr., eased the fireboat inside the V-shaped pocket of flames where the ships were impaled. Continuing explosions aboard the *Esso Brussels* and ominous rumblings warned of possible eruptions which could destroy the *Fire Fighter* and kill its crew.

As worrisome were some 300 blazing cargo containers stacked on the *Sea Witch*'s main deck. Never before had the American fire service been confronted with a containership fire of this magnitude. Were hazardous materials inside those cargo boxes? Would the steel lashdowns snap and send dozens of heavy cargo containers tumbling down onto the *Fire Fighter*? While all of the cargo containers burned, along with their combustible contents, none held dangerous chemicals and none broke loose.

Flames and gobs of smoke boiling from both ships led the *Fire Fighter*'s crew to doubt that anybody was still alive aboard either ship. Their surprise at finding two huge vessels crunched together was matched only by the sight of someone with a flashlight on the fantail of the *Sea Witch*. Fitzsimmons edged the *Fire Fighter* forward as heat scorched the bow and turned the fireboat's metal foredeck hot. Still the *Fire Fighter*'s crew, enduring incredible heat and smoke, remained at their deck pipes. Then they saw that there was not one crewman signaling, but dozens huddling around him while desperately hoping for escape from the withering flames and overwhelming smoke.

Normal procedure would have required mooring of the fireboat against the *Sea Witch*, so both vessels would be in rhythm as rescues were attempted. There was no time for that. As Fitzsimmons held the *Fire Fighter*'s bow against the *Sea Witch*, other crewmen raised a boarding ladder. The fireboat firefighters assisted 31 *Sea Witch* crewmen down to the *Fire Fighter*'s deck. Paterson's body also was brought on board.

Assured there were no other survivors, the *Fire Fighter* raced to shore while its crew attended to severe burns suffered by the crew of the *Sea Witch*. Off-loading the survivors at Brooklyn's 69th Street Pier, where a FDNY Disaster Unit awaited, the *Fire Fighter* raced back to the battle which was now being fought by other FDNY fireboats and other vessels while U.S. Coast Guard and other launches searched for survivors.

By dawn, the interlocked *Esso Brussels* and *Sea Witch* drifted aground in an area appropriately named Gravesend Bay. The tanker fire was largely controlled by dawn, but the *Sea Witch*'s containers continued to burn. The vessels were pulled apart around 6:30 by tugs from Moran, McAllister and other towing companies in delicate maneuverings which were not without their own extreme dangers. The *Sea Witch* was beached at Norton Point in Gravesend Bay, about 1500 yards from the *Esso Brussels*.

Firefighting continued for two weeks as fireboat and land-based firemen went aboard the *Sea Witch* to battle residual flames below deck. Altogether, 74 crewmen of both ships were saved, but 17 died. Ship and cargo losses came to $23 million.

Citations were awarded to crews of the many large and small boats which participated in rescues following one of New York's worst shipboard disasters. The *Fire Fighter* and its crew received the 1974 American Merchant Marine Seamanship Trophy for distinguished seamanship and heroism.

On May 22, 1975, the *Fire Fighter* and its crew were presented with the U.S. Department of Commerce Gallant Ship Award, the highest commendation the federal government can make to a merchant vessel. The Gallant Ship bronze medallion is mounted on a bulkhead in the *Fire Fighter*'s Gold Room, where nozzles and other fittings are kept. The medallion depicts a ship steaming at full speed. Beneath is a description of the gallantry of the *Fire Fighter*'s crew. When the Gallant Ship award was made, Lieut. James F. McKenna of the *Fire Fighter* received the Merchant Marine Meritorious Service Medal. The nine other crewmen aboard the *Fire Fighter* during its unprecedented feat, were presented with Gallant Ship Unit Citations and Ribbon Bars.

The era of supertankers forced construction of special facilities to accommodate these deep-draft vessels which usually could not be guided safely into berths inside ports. There were, moreover, other considerations. Fire protection authorities put it this way: The bigger the boat the bigger the boom if it explodes. All the more reason, therefore, to build supertanker facilities in outer harbor areas where the disaster potential would be somewhat minimized.

The Port of Los Angeles Supertanker Terminal extends from Berths 45-47 on a bulge of land spearing into the Outer Channel. The area is fairly well removed from major industrial complexes and residential districts. On Friday night, December 17, 1976, the Liberian registered *Sansinena* stood at Berth 46. The 810-foot-long tanker, largest of its kind when built in 1958, had unloaded more than 20 million

*The Gallant Ship Award, the highest commendation the U.S. Government can make to a merchant vessel, was presented to New York's **Fire Fighter** fireboat in recognition of the crew's heroism while rescuing 31 crewmen from the fantail of the blazing **Sea Witch** shown after it was beached. Frank Duffy*

*San Francisco's Pier 20 on The Embarcadero survived the 1906 earthquake and fire, but was destroyed during a five-alarm fire, November 30, 1972. The 9600 gpm* **Phoenix** *and 35 land-based fire companies successfully prevented the condemned pier fire from spreading. From the Collection of Paul Ditzel*

Although night winds were calm and there was barely a ripple on the Delaware River, thermal drafts resulted from the intensity of a six-alarm fire, October 30, 1974, which devastated 23 acres inside the Reading Railroad's Port Richmond Yards, including three gigantic piers. The thermal drafts bent the 6000 gpm

*Bernard Samuel's* streams and created water turbulence. The fire, seen for 40 miles, also destroyed warehouses, melted railroad freight cars and ignited carloads of coal. Philadelphia fireboats shot protective water curtains and prevented ignition of a nearby grain elevator. From the Collection of Bob Burns

Underwater thruster nozzles hold Los Angeles' **Ralph J. Scott** alongside the **Nortrans Visions**, loaded with burning metal turnings. The fireboat's lift basket was effectively used, May 2, 1974, to raise firemen and hoselines to the freighter's deck. From the Collection of Bill Dahlquist

**Fireboat No. 110** of the Los Angeles County Fire Department protects Marina del Rey, the largest man-made small craft harbor in the world. The fireboat also would respond to nearby Los Angeles International Airport if a plane crashed into the Pacific Ocean. The 1500 gpm at 150 psi craft was put in service in 1976 and features an aluminum hull. Its specifications include a length of 37-feet, a beam of 14-feet, 9-inches and a 3-foot draft. Displacing 13 tons, **Fireboat No. 110's** propulsion plant includes two 3208 Caterpillar diesel engines. A 3160 Caterpillar drives the single-stage Waterous pump. The Stang Intelligiant bow monitor delivers 1250 gpm. Crewing the boat are a captain, a marine engineer and two firemen. Additional personnel are quickly available from the land-based station where **Fireboat No. 110** is berthed. From the Collection of Paul Ditzel

America's most famous fireboat station is Pier A, Battery Park, at the southern tip of Manhattan Island and serves as headquarters of the FDNY Marine Division. The 285-foot-long pier was placed on the National Register of Historic Places by the National Park Service on June 27, 1975. The pier, completed in May, 1886, features a four-story clock tower. The clock, donated by Daniel G. Reid, a

founder of the U.S. Steel Corporation, strikes the hours in ship bells and is one of few of its kind. The FDNY Marine Division took full use of the pier in 1960. The 19,000 gpm **William D. McKean**, flagship of the Marine Division fleet, is berthed at Pier A. Frank Duffy

111

Thousands of mushroom-shaped rivets were sheared from the oil tanker *Sansinena* when it exploded, December 17, 1976, at the Port of Los Angeles supertanker terminal. The rivets pelted the waterfront for miles and plinked into harbor waters. From the Collection of Tony DiDomenico

Boston's newest and fastest fireboat went in service September 15, 1976. With a top speed of 22 knots, the **St. Florian** has a pumping capacity of 3000 gpm and is 45-feet-long, has a beam of 15-feet, 10-inches and draws 4-feet-2-inches. The boat also serves as a launching platform for SCUBA divers. Bill Noonan

The bow section of the tanker, **Sansinena**, juts above the Los Angeles supertanker terminal following massive explosions, December 17, 1976. Los Angeles fireboats **Bethel F. Gifford** (left) and **Fireboat No. 4** assist with fire attack operations while U.S. Coast Guard and other small craft search for survivors. Phil McBride

gallons of Indonesian crude oil for the Union Oil Company. The *Sansinena* was taking on fuel and ballast for a scheduled 11 p.m. departure.

An empty tanker can be as hazardous as one that is loaded. The *Sansinena*'s 36 cargo compartments were filled with extremely explosive hydrocarbon vapors that remained after the crude oil was off-loaded. Safety protocols and built-in shipboard equipment is supposed to protect against vapor explosions. In the case of the *Sansinena*, this equipment was poorly designed. Compounding the probability of explosions was poor safety feature maintenance, thus permitting many vapor leakages.

Shortly after 7 p.m., while many Christmas parties were starting in waterfront restaurants and residential areas, a massive vapor cloud began to form between the midship and the after deckhouse of the *Sansinena*. The slight evening breeze was not strong enough to dissipate the cloud which ballooned while seeking ignition. At 7:38 p.m., a flashing explosion, probably near the midship deckhouse, was seen for miles. A second and more powerful explosion instantaneously followed.

The explosions lifted a 500-foot-long midship section, weighing thousands of tons, straight up. Mathematic computations and witness accounts, including those of airline pilots, showed this gigantic chunk of the *Sansinena* soared 750 feet high before whanging down onto the supertanker terminal. The mammoth wad of iron crushed buildings and severed oil pipelines which freely gushed.

The explosions were heard for miles. Concussions shattered waterfront restaurant windows and caused damage in a six mile radius. Tens of thousands of the *Sansinena*'s rivets, about the size of large mushrooms, pelted the area and

The **Ralph J. Scott** stands alongside the after section of the shattered **Sansinena** oil tanker following explosions which killed six crewmen and three supertanker terminal workers. A 500-foot-long midship and deckhouse section of the **Sansinena** continues to smoke where it slammed down onto the terminal at the southern end of the Port of Los Angeles. The explosions lifted the section 750 feet over the terminal before the massive chunk of the tanker plummeted onto the terminal. The three firemen of **Fireboat No. 5** were awarded Medals of Valor for rescuing 17 **Sansinena** crewmen who jumped or slid down a line at the port stern of the tankship. From the Collection of Bill Dahlquist

plinked into port waters. Force of the explosions pushed the severed bow and stern sections 60 feet from the dock which was heavily damaged. Some 12,000 gallons of the *Sansinena*'s blazing bunker fuel sent columns of black smoke high over the fire lighting the southern tip of the Port of Los Angeles.

At the Berth 29 station of *Fireboat No. 5*, Supervising Mate Walter L. Ball was dumped from his office chair by the blasts. With Firefighters John Kemperman and Forrest E. Taylor, he hurried aboard the small fire-rescue boat. Debris pelting the fireboat had shattered the flying bridge windshield and cracked a starboard cabin window. Along with other fireboat and land-based firemen, *Fireboat No. 5* started for the glare even as fire alarm switchboards lit up with calls.

It was but a short dash of about 1000 feet as the crew of *Fireboat No. 5* saw smoke and flames boiling from the arrow-shaped hulk of what remained of the *Sansinena*'s bow. Hearing screams, Ball brought the fireboat to the stern where the firemen saw trapped crewmen who had thrown a rope overboard. Some were preparing to slide down it, while two others had already jumped into the oil-slickened water. Ominously loud hissing and eerie whines from inside the stern section only added to the crew's panic as they braced for still more explosions.

Firemen Kemperman and Taylor pulled the crewmen in the water on board the fireboat, while Ball held the bow against the stern. Heat raised paint blisters on the bow as the firemen helped the crewmen as they slid down the rope and onto the fireboat. Others jumped onto the fireboat and suffered fractures. The fireboat firemen saw the tide was pushing floating flames toward them. Ball could not activate the bow monitor gun to push the fire away out of concern that the stream would hit the crewmen sliding down the rope or jumping into the fireboat.

The fireboat's crew recognized a new danger as the small craft began taking on water and settled steadily lower from the weight of 18 *Sansinena* crewmen and the three firemen. The tanker's master was the last down the rope. Assuring firemen that there were no other survivors aboard, Ball headed for the LAFD's Paramedic Medical Command Post at Berth 55.

Returning to the *Sansinena*, the fireboat firemen joined the crews of other fireboats, including the *Bethel F. Gifford* and the *Ralph J. Scott*, which were fighting the flames with water and foam while small craft searched for survivors. Land companies of firemen, meanwhile, found footing treacherous as they tried to make their way into the terminal. The oil slick from the ruptured pipelines forced them to lay hoselines by hand. Their difficult job was complicated as the lines snagged on twisted hunks of steel flung by the explosions. Later that night, land and fireboat companies controlled the fire.

The *Sansinena* catastrophe demonstrated the wise choice of a fairly remote location for a supertanker terminal. Damage was largely confined to the immediate area of the terminal and came to $21.6 million. Six *Sansinena* crewmen were killed as were three supertanker terminal employees. The 18 crewmen rescued by *Fireboat No. 5* recovered from their injuries.

*The City of Miami* was the nation's first amphibious fireboat. Miami put the 2500 gpm boat in service in 1979. The city chose an amphibian in the belief that the 70-foot-long craft could more quickly reach waterfront alarms by taking short cuts across city streets. From the Collection of Paul Ditzel

Boston Whalers Inc., of Boston is a leading supplier of small fireboats. Included in their model line is the Guardian with a 700 gpm at 100 psi pumping capacity. The 24-foot, 7-inch boat attains speeds of up to 45 miles per hour. Other specifications include an 8-foot beam and a 16-inch draft. The Guardian is equipped with a 4-cylinder gasoline-powered pump engine and a Darley KSW 1250 bronze pump. It is propelled by dual Johnston V6 150 horsepower outboard motors. Other features include a bow boarding platform and a towing rail tower. Boston Whaler, Inc.

Few of those attending the June 14, 1978, LAFD Medal of Valor luncheon had been on the department long enough to recall the unrecognized heroism of fireboat firefighters during the 1944 and 1947 *Fredericksburg* and 1947 *Markay* disasters. For the first time in the 92-year history of the LAFD, fireboat firefighters were at last recognized for what had heretofore been described as "only doing the job they get paid to do." For gallantry while rescuing the 18 *Sansinena* crewmen, the LAFD's highest award, The Medal of Valor, was presented to Supervising Mate Ball and Firefighters Kemperman and Taylor.

From the late 1970s onward, many innovations began to change the traditional look of fireboats. In 1979, Miami put in service a 70-foot-long amphibious fireboat, *City of Miami*, with a 2500 gpm at 150 psi capacity.

The vehicle was intended to more quickly reach water-front alarms by traversing city streets to port locations where it entered the water. Milwaukee, in 1984, followed suit with the *Amphibian* of similar size and pumping capacity. Both boats were seen as a cost-effective means of using fireboats to perform similar land and water services.

From the Pacific Northwest, however, came more successful approaches to long-range port protection problems. The neighboring cities of Seattle and Tacoma provided examples of future trends in fireboat design and construction by purchasing craft which are not only dissimilar in appearance, but evidence what fireboat history has demonstrated: That Utopian day when there is a standard piece of apparatus — land or water — so long bruited within the fire service — is no closer as the 21st Century nears than it was when fire engines were pulled by horses.

As closely parallel and as geographically close as Seattle's and Tacoma's waterfront protection requirements are, a comparison of their boats points up the fire service tradition of different approaches to similar problems.

If ever a fireboat was built for current and projected needs of a port, the *Chief Seattle* is a prime example. From an historic perspective starting in the early 1900s and continuing for nearly 75 years, Seattle's waterfront was protected by the 22,800 gpm *Duwamish*, built in 1909 and the 1927-built *Alki*, which added another 18,000 gpm to the fleet's capabilities. Those were the days when Seattle's waterfront was a patchwork of wooden docks, lumber yards and mills, cedar shake shingle factories, wood-framed buildings containing unprotected combustibles and wooden vessels.

By 1980, the Port of Seattle had undergone significant changes. Largely replacing the old conflagration breeders along the city's 75 miles of waterfront were concrete docks, fire-resistant structures with protective water sprinkler systems and recreational facilities, including marinas for hundreds of pleasure boats. Not once during the 20 years preceding the early 1980s had a waterfront fire required a fireboat discharge of more than 7500 gpm.

Seattle decided it would retire its *Duwamish* after acquiring a new high-speed, multi-purpose fireboat of 8500 gpm

*Seattle's old 18,000* **Alki** *fireboat, built in 1927 (right) salutes the new 8500 gpm* **Chief Seattle***, acquired in 1984. In a repeat of Seattle's fireboat history, both were designed by the same firm which, by 1984, operated under the name of Nickum & Spaulding Associates, Inc. From the Collection of the Elliott Bay Design Group*

*Ice clings to Boston's **Fire Fighter** during a six-alarm pier fire which taxed Boston firefighters for several days, starting January 20, 1984. The **Fire Fighter**, put in service June 7, 1972, is outfitted with two General Motors 12V-71N diesel engines; equally divided for propulsion and pumping. The boat's twin A0818L De Laval pumps deliver 6000 gpm at 150 psi. Specifications include: 76-feet overall length, a beam of 19½-feet and a draft of 5½-feet. The **Fire Fighter** is equipped with four 4-inch diameter Stang Intelligiant guns. As built, the boat was equipped with an articulated elevating water tower mounting a 2000 gpm gun. Mechanical breakdowns led to its removal. Bill Noonan*

capacity and designed for cost-effective operation by a three-man crew. The resulting fireboat was a blend of proven concepts and high technology in this age of computer modeling.

History was recalled when the design contract for the $2.2 million boat went to Nickum & Spaulding Associates, Inc., whose predecessor, W.C. Nickum & Sons Company, Inc., designed the *Alki*. The design team included Nickum & Spaulding's Paul A. Gow, project manager and Edward C. Hageman, chief naval architect, along with Captain (later Deputy Chief) Richard A. Columbi of the Seattle Fire Department who oversaw the project for the city.

To fulfill Seattle's criteria for a high-speed, shallow draft boat with a 40-year life expectancy, Nickum melded its fireboat experience with computerized library research and computer-drawn isometrics. The result was a boat with a V-contoured planing hull of welded aluminum. Calculations confirmed that the design of this contoured hull would meet requirements for a 26-knot fireboat which would create safe and rapidly-dispersing wake patterns with crests of only about 30-inches at full speed. Lightweight aluminum enabled the use of less costly engine room machinery without diminishing operational horsepower requirements.

*"Old Ironsides", the frigate **USS Constitution**, permanently berthed in Boston, makes its annual voyage, July 4, 1985. The trip is made to turn the vessel so it is evenly weathered on the port and starboard sides while docked. Among the boats escorting "Old Ironsides" is the fireboat, **Fire Fighter**. Bill Noonan*

Nickum & Spaulding designed a boat 96½-feet-long with a beam at main deck of 23-feet and a draft of only 7-feet. The craft was constructed at Nichols Brothers Boat Builders, Inc., Freeland, on Whidbey Island, Wa., north of Seattle. Christened *Chief Seattle* on August 17, 1984 — the name was chosen by Seattle elementary school children who would be rewarded with a ride aboard her — the fireboat in profile more closely resembles a large luxury yacht than it does a traditional fireboat.

The *Chief Seattle* has three Detroit 16V92TI Marine diesel main propulsion engines, each developing 1000 horsepower at 2300 rpm. They drive stainless steel, four-bladed Coolidge propellers. Electrohydraulic steering systems and controls are operated from the pilothouse. The boat, with its fully-automated engine room, was designed to obviate the necessity of stationing marine engineers below deck to supervise and sometimes operate propulsion and pumping machinery. Electric power is mainly supplied by John Deere 6-cylinder, 40-kilowatt generators.

The pilothouse atop the deckhouse insures maximum 360-degree visibility for the pilot and boat commander. Maneuverability is enhanced by four 2000 gpm thruster nozzles located near the waterline on the port and starboard sides of the hull. The two forward thrusters also serve underdock firefighting needs. The pilot, utilizing combinations of propeller speeds and thruster nozzle outputs, can, with the triple-screw and rudder configurations, hold the *Chief Seattle* on station.

Three Worthington 8LR20 single-stage centrifugal pumps are each rated at 2500 gpm at 150 psi. The pumps

*Seattle's skyline forms a backdrop during sea trials of the **Harry Newell**, a 5000 gpm at 150 psi fireboat delivered in 1986 to the Ketchikan, Alaska, Fire Department. Named after the only Ketchikan fireman to die in the line of duty, the **Newell**'s 30-knot top speed was intended for fast coverage of the city which is sometimes described as five miles long and two blocks wide. The **Newell**, designed by Naval Architect Jack Sarin and built by Workboats Northwest, Inc., Seattle, is powered by two Detroit Allision 6-71TI turbocharged diesel engines, each developing 410 horsepower. The aluminum alloy welded boat's four 1250 gpm American pumps were supplied by United Fire Service. Workboats Northwest, Inc.*

*Oakland's fireboat, **City of Oakland**, the former **Hoga** of Pearl Harbor fame, cools the hull of the burning oil tanker **Porto Rican**, which exploded and burned early October 31, 1986, outside of the Golden Gate. Joining the **City of Oakland** in the battle are tugs from the Treasure Island Naval Station, while helicopters lift off survivors. One tanker crewman was killed and 28 were rescued. The U.S. Coast Guard commended the **City of Oakland** and her crew for their efforts. From the Collection of Michael J. Cline*

are driven, via twin-disc Omega reduction gearing and Systems Engineering clutches, by the main engines without significant detriment to either pumping or maneuvering capabilities. Noise mufflers in the engine room and pilothouse provide what Gow described as "a particularly mellow tone" as contrasted to the usually expected engine room and pilothouse din which, in other boats, often leads to permanent hearing impairment among fireboat firefighters.

In addition to the two underdock nozzles, the *Chief Seattle* has four firefighting monitors which are mostly remote-controlled from the pilothouse. The main Stang-built monitor is tower-mounted atop the pilothouse and rated at 5400 gpm when fitted with an 8-inch diameter nozzle, or dispersal of 500 gpm of foam. The *Chief Seattle*'s second largest gun is the 4400 gpm Stang Shaper with a 6-inch nozzle. The only wholly-manually operated monitors are the two 2500 gpm, 4-inch diameter Stang Shapers mounted at the aft end of the bridge deck.

The tower monitor, probably unique in the American fire service, is raised by water pressure to a height of 45-feet above the waterline. The idea originated around 1898 with

Captain Henry J. Gorter, master mechanic and superintendent of steam fire engines for the San Francisco Fire Department. He used water pressure to raise the horse-drawn, land-based, water tower apparatus he built for San Francisco and other cities along the Pacific Coast, notably Los Angeles. The tower can be remotely-controlled or manually-operated.

Extra firefighting capability is offered by 14 gated 3 1/2-inch hose outlets evenly distributed on the port and starboard sides of the deckhouse. Automatic proportioners by National Foam supply the tower monitor or port and starboard deckhouse discharge outlets for coupling to standard 2 1/2-inch diameter handheld hoselines. The *Chief Seattle* is protected from radiant heat by a downward-directed water spray system mounted around the bridge and main deck areas.

As recent experience has shown, fireboats are far more frequently used for emergency search and rescue incidents than major waterfront blazes. Sensitive to the growing multiplicity of fireboat uses, the full width of the *Chief Seattle*'s stern features a water-level platform with easy access from the main deck. This sloping ramp, an integral part of the

Long Beach, Ca., purchased two identical 10,000 gpm at 165 psi fireboats, **Challenger** and **Liberty**, which were delivered in 1987 and 1988. The boats, rated at 15½ knots, were designed to reach any port alarm in 10 minutes, including two minutes of startup time. Because Long Beach has many offshore oil drilling platforms, the boats are specially equipped for deep water service. Their 64-foot telescopic towers can deliver foam or water on burning oil rigs or supertankers. The **Challenger** and **Liberty** each carry 1000 gallons of foam concentrate. Inset photo (upper right) shows a SKUM blabbermouth-type nozzle for delivering straight or fan-like spray streams as well as foam dispersal. The boats are virtually completely remote-controlled and have fully automated engine rooms. Costing $5.4 million, the boats were designed by Nickum & Spaulding Associates, Seattle, and built at Moss Port Marine, Escatawpa, Miss. The boats have welded steel hulls and aluminum superstructures. Other specifications include their overall length of 88½-feet, a beam of 21-feet and a 6-foot draft. Salt water corrosion and other problems that occurred soon after delivery were being addressed in 1989 by modifications costing around $1 million. From the Collection of Bill Dahlquist

hull, facilitates water rescues or off-loading of a small Boston Whaler motorboat for retrieval of victims. The ramp directly leads to an emergency medical services treatment room in the after end of the deckhouse. When fully-equipped, paramedics will be able to simultaneously aid two immobilized trauma patients and other injured persons. Built-in cabinets hold medical supplies, pre-heated blankets and other necessities for treating hypothermia victims.

While operating in the rescue mode, the pilot can maneuver the *Chief Seattle* from the remote station on the aft end of the bridge deck, thus providing him full view of delicate rescue operations requiring close operational coordination of the fireboat with the rescuers and their patients.

As police departments and the U.S. Coast Guard rapidly downgrade their harbor security and patrol activities, the *Chief Seattle* was designed to fill these functions. The centerline engine is used for 5-6 knot patrol activities at low operational costs and for sustained periods. The *Chief Seattle* carries 1425 gallons of diesel fuel.

Further indicative of late 1980 trends in fireboats, the *Chief Seattle* is expected, by the end of 1989, to become equipped with a computer and Cameo Program developed by the National Oceanic and Atmospheric Administration. As a floating command post, chief officers and others will be able to instantly access, via video display terminals, information concerning thousands of hazardous materials commonly found aboard ships or in waterfront industrial and cargo container complexes.

No longer will incident commanders be forced to place telephone calls throughout the United States and sometimes abroad, to determine flammability, toxicity, medical treatment and appropriate firefighting protocols for the myriad of hazardous materials in international commerce. Data banks aboard the *Chief Seattle* will be able to provide this information for command and control of emergency situations.

Zeroing in on specific Seattle waterfront target hazards, the *Chief Seattle*'s computer system will have the instantaneous video display capability showing diagrams of piers, along with detailed information on its structures, water supplies and other instantly-needed information to safely and effectively cope with problems ranging from chemical spills to full-blown fires.

Tacoma, on the other hand, advanced fireboat design and technology dramatically further when, on December 24, 1982, it put in service its futuristic *Defiance*, the United States' first catamaran-like fireboat and, several months later, December 28, 1983, followed it with the identical *Commencement*. The boats were named after local landmarks: Point Defiance and Commencement Bay.

Each of these surface-effect-system (SES) fireboats skims the surface of port waters while riding on bubbles of air trapped inside an envelope formed by the downdraft created by fans of a diesel blower engine. Bubble containment is by rigid port and starboard longitudinal sidewalls that extend into the water as well as flexible bow and stern skirts. This flexibility not only reduces normal drag caused by the water, but helps to absorb the impact of unseen underwater objects. This feature paid off when one of the fiberglass-reinforced-plastic hulled boats struck a submerged barge. Damage was quickly repaired and the boat returned to service.

Perhaps the best insight into the operation of these SES fireboats is to take a hypothetical ride aboard one of these sleek, snub-nosed boats that look like barges. Looks are deceiving, because these multiple purpose boats were built for speed, maneuverability and the many uses which

*The **John D. McKean** fireboat spouts welcoming red, white and blue-colored streams while Moran Towing Corporation tugs assist the **Crown Odyssey** past the Statue of Liberty during the cruise liner's maiden arrival in New York, September 13, 1988. The multi-deck **Crown Odyssey**'s passengers are afforded picture window views, rather than those from traditional portholes. Frank Duffy*

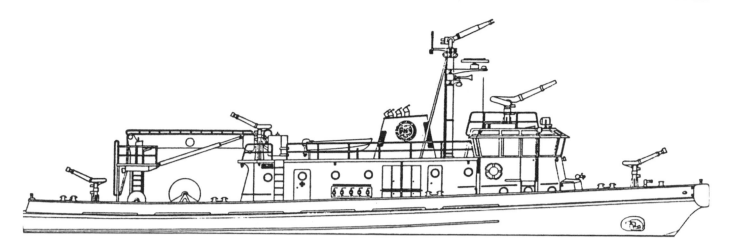

Los Angeles, planning for port protection problems through the year 2020, is expected to add a new fireboat to its fleet during the 1990s. This artistic representation will most likely result in a 15,000 gpm fireboat, 100-feet-long, with a beam of around 23-feet and a draft of 7-feet, 8-inches. Cognizant of increasing petrochemical hazards in the rapidly-expanding port, the boat will likely carry 2000 gallons of foam concentrate. Projections call for seven engines, three fire pumps, nine monitors, including one mounted on a 60-foot extendable tower and a 60-foot lift basket, together with 16 hose outlets. Also configured into the planning will be an emergency medical service treatment area as well as a command-control center with appropriate video display terminal access to decision-making protocols ranging from hazardous materials to specific areas of the port. From the Collection of Paul Ditzel

With snow-capped Mount Rainier in the background, Tacoma's **Defiance**, America's first cataraman-type fireboat, skims across Commencement Bay while traveling upon an enclosed bubble of air created by the downdraft of five diesel-operated fans. TFD

Pilot Jim Nelson at the starboard control panel of one of Tacoma's two surface-effect-system fireboats which may revolutionize fireboat design and construction into the 21st Century. The boats were designed for remote control operation by the pilot and a captain. TFD

today's and tomorrow's fireboats must meet to justify their costs. In addition to firefighting, these include fire prevention patrol, search and rescue, water pollution monitoring, towing of disabled craft, harbor security patrol and dewatering of the holds of flooded vessels.

Historic innovation is sensed when you notice the pilot sits, not stands, at the starboard side remote control propulsion and steering panel which resembles that of a yacht. Beside him is the boat's only other necessary crewman: the officer who remotely controls pumps, the four 1500-5500 gpm monitors, two underwharf firefighting nozzles and the telescoping elevating ladder. If needed, additional personnel are provided by land-based fire companies. Although each boat was, by 1989, crewed by three firemen under labor contracts, the recollection is inescapable: During the 20th Century, large fireboats sometimes required crews of 15 and it was that expense which sent many fireboats to early retirement or reserve standby status.

Leaving the dock, the pilot rapidly accelerates the twin stern-mounted diesel propulsion engines. At about 15 knots, he engages the forward air blower engine. The downdraft of the engine's five fans causes the fireboat to rise about 3 1/2-feet above the surface of the water while its sidewalls remained immersed. The sensation of zipping along on air bubbles is akin to what airline advertisements used to refer to as gliding on a cloud. Even in fairly rough swells of six feet, the boat bubbles along on a virtually horizontal plane while easily kissing the crests of the waves.

In the usually smooth waters of Commencement Bay, Tacoma's SES fireboats can achieve speeds of around 30 knots. Swells only slow the boat to about 20 knots as the smooth-as-silk ride continues. Whisking across Commencement Bay, a backward glance notes that there is no potentially dangerous wake which could damage small craft.

Maneuverability and quick response to controls are other features of Tacoma's fireboats. Catamaran sidewalls, along with quick control of the twin under-the-hull rudders, vir-

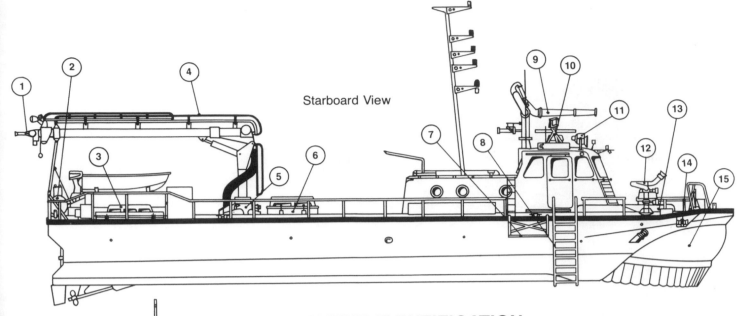

Starboard View

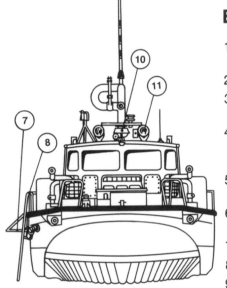

Bow View

## EQUIPMENT IDENTIFICATION

1. 1500 GPM Remote Control F/S* Nozzle.
2. Foam Tank.
3. Propulsion Engine Compartment Hatches.
4. 35' Telescoping Ladder and Waterway on Crane with 2800 pound lifing capacity.
5. 7075 GPM Manifolds w/2½" and 5" hose connections.
6. Main Fire Pump Engine Compartment Hatch.
7. Rescue Platform (detachable).
8. Divers Ladder (detachable).
9. 7075 GPM Remote Control Nozzle: 90 degree rotation   −15 degree to +70 degree Elevation.

10. High Definition Radar.
11. Searchlight and Loudhailer w/¼ mile listening device.
12. 2500 GPM Remote Control F/S* Nozzle: 270 degree rotation w/stops.
13. Forward Lift and Forward Fire Pump Engine Compartment Hatch.
14. 2500 GPM Remote Control Under Dock F/S* Nozzle: −5 degrees +10 degrees Elevation.
15. Bow Skirt.

*F/S = 90 degree Fog to Straight Stream.

NOTE: Foam delivered thru 1½" deck hyrants at base of Ladder and thru Ladder Nozzle.

tually prevents the traditional tendency of side slippage or skidding at high speeds. The pilot can, moreover, turn the boat in a full circle in virtually the 70-foot-length of the craft and, at full speed, bring it to an emergency stop in about 200 feet.

Arriving at the fire or other emergency, the air blower is disengaged. The SES boat settles into the water to provide an operating platform while drawing only 5-feet, 4-inches. The sidewalls result in remarkable stability as the fireboat becomes a 1400-square-foot platform. The boats' low free-board and detachable rescue ladders give quick access to and retrieval of persons in the water, as well as facilitating SCUBA operations. In firefighting mode, the SES boats can operate their monitors alongside piers or boats with result-ing lists of less than 2 per cent and minimal back water pres-sure problems.

From concept to delivery of America's first SES fireboats spanned a decade marked by intense research, a willing-ness to consider new ideas which would shatter long-held shibboleths and a skepticism so profound that it could have destroyed the credibility and careers of Tacoma fire officials who espoused SES ideas.

Tacoma's revolutionary fireboats trace their lineage to the 1929-built Tacoma *Fireboat No. 1*, which was originally gaso-line-powered, steel-hulled and with a pumping capacity of 10,000 gpm. The 96½-foot-long boat was suited for a Tacoma of an earlier era when protecting the city's major lumbering industry, wooden piers and warehouses was of paramount concern.

By the 1970's, Tacoma's needs had changed as dramati-cally as the SES boats which would replace *Fireboat No. 1*. Just as other cities normally did, Tacoma's approach to mod-

The success of Tacoma's innovative surface-effect-system fireboats led to the 1982 retirement of **Fireboat No. 1**, after 53 years of service. The waterfront landmark, permanently displayed in the city's Marine Park, sits inside a dry berth resembling a graving dock. George Thornhill

Tacoma's **Commencement**, like its sister fireboat, **Defiance**, delivers more than 7000 gpm. Monitors are remotely-controlled from the pilothouse by the captain. TFD

ernism was typical: Rebuild *Fireboat No. 1*. The $1 million cost was not only unacceptable, but would not result in a boat fulfilling the port's burgeoning requirements. By 1989, the Port of Tacoma's 38 mile waterfront would include commercial, residential and recreational areas. Its 13 miles of industrial waterways necessitated more than a rebuilt, sluggishly slow boat that already had over 40 years of service.

Choice of SES fireboats was not a whimsical decision. Long before Tacoma opted for SES craft, the fire department studied fireboat protection in 17 American cities. Ultimately, the new fireboat team included Fire Chief James W. Reiser, Deputy Chief (later Chief) Tony Mitchell, who had a dual responsibility as harbormaster; Stan Mork, deputy

chief harbormaster, Consulting Engineer John Kuhn, a systems analyst and John Maddock, Tacoma Fire Department administrator of research and development. Playing a lead role in the study was Owen Douglas, who had a depth of maritime expertise in Hovercraft vessels.

Parameters for Tacoma's needs were developed over a nine year period, including results of a study started in 1973 and funded by the U.S. Maritime Administration. Publication of these results contained 42 guideposts which were not only germane to Tacoma's specific needs, but were expected to seed future thinking of fireboat designers and builders. Among the report's major recommendations for what could be described as a generic fireboat were:

1. Operational by a two-man crew.

2. A dash speed of 30 knots and excellent rough sea performance in up to 6-foot waves.

3. Shallow draft and low wake.

4. A hull lifespan expectancy of 20 years.

5. Capabilities for pumping 5500 gpm for eight hours.

6. Multiple fixed monitors for firefighting.

7. Superb maneuverability under all operational conditions.

8. Capability of on-deck access to large vessels.

9. Excellent rescue equipment and related facilities.

Synthesizing their expertise, research and other resources, the Tacoma team decided upon acquisition of two SES fireboats whose cost would have to be held to $2.3 million from the city. The SES decision understandably rattled the fire service to its very bedrock of conservatism. A century of experience held steadfast to the idea that fireboats had to be heavy-duty and similar in profile to the tugboats from which they had evolved.

Building large fireboats with hulls of fiberglass-reinforced-plastic might be acceptable in small fire-rescue craft, but was anathema to those who were wedded to the idea that fireboats had to be of reinforced steel construction and powered by mighty diesel-electric machinery to drive heavy-duty fire pumps.

So ludicrous seemed the idea that an excellent fireboat could be built of synthetics and travel on an air bubble, that Tacoma's fireboat design team came in for ridicule locally and nationally, especially Chief Tony Mitchell who became the leading spokesperson for the new concept. In deference to critics, it must be noted that the SES concept had never proved itself in the United States fire service. As history has shown, most new ideas must be tried and found workable by a fire department before other departments are likely to take the gamble themselves, especially when so much money and property to be protected is at stake. Long before they were seen in Tacoma, critics dubbed SES fireboats "Tony's Follies" and "Tony's Toys".

Controversy was further fueled when Tacoma called for construction bids and inserted another factor which flew in the face of traditional apparatus acquisitions. Standard procedure to guide prospective bidders are specifications. These specifications are routinely noted in requests for proposals to build fireboats or any other type of fire apparatus.

Tacoma, with $2.3 million to spend and not a penny more, took a different course. In still another break with precedent, builders were put on notice that performance would be prioritized and design specifications secondarily. Mitchell gave this explanation for the turnabout:

"The typical method of having a desired product built or manufactured is to develop a design specification and then award a contract to the lowest bidder. With this method, what you tell the builder is generally what you will receive. But (the finished product) may not measure up to all your expectations," as fireboat history has often demonstrated.

Mitchell expanded upon the rationale for the bidding procedure in a statement which could well have long-range impact upon fireboat construction: "Because the product we were pursuing was unique and nonexistent and because we had limited resources which did not include the developmental cost, we opted for primarily a performance specification. Simply put, if the craft did not perform according to our strictest standards, then we would not pay the builder."

Virtually overlooked in this brouhaha was the fact that the idea of air bubbles serving as a lubricant or surfactant to reduce surface drag of vessels is about a century old and thus pre-dates every large fireboat in service in the United States.

In the late 19th Century, naval architects were intrigued by the concept of lifting boats out of the water by creating air bubbles. The idea offered many paybacks. By traveling on a pocket of air, boats would be less expensive to build because lighter weight and therefore cheaper propulsion machinery could be utilized. Without the significant water drag of conventional vessels, boats would be faster, more maneuverable and create a highly beneficial domino effect, including lower labor and maintenance costs.

Before the mid-20th Century, air lubrication travel was only a theory because hulls, notably those of fireboats, were of heavy wood or metal. Machinery to lift them by air out of the water was not technologically practical. In 1950, however, Sir Christopher Cockerell invented an air cushion system which efficiently trapped a bubble on the underside of vessels. This ground-effect-machine (GEM) invention became feasible for marine applications starting around 1966 with the development of rugged, lightweight hull materials and designs.

Vosper Hovermarine of Southampton, England, pioneered the design and construction of Hovercraft, mostly for passenger commuters in trans-channel service in England, the North Sea and the Orient. As Mitchell explains, there is a significant difference between GEM craft and SES boats. GEMS can best be compared to amphibious aircraft. SES fireboats are more directly related to amphibious ships. "SES fireboats offer higher speeds (than conventional fireboats) with lower (cost) installed horsepower than equivalent planing hull type fireboats."

Vosper Hovermarine was the only bidder when Tacoma called for proposals. In retrospect, the question is probably moot whether American designers and builders shunned the project because they were less than enthusiastic about the unorthodox idea of "If they don't work, we won't pay for them" or, more likely, that they simply lacked SES expertise. Whatever the case, the contract for the two boats, each weighing 39 tons, went to Hovercraft.

Controversy flared afresh, as might be expected, given the Pacific Northwest's historic reputation for excellence in shipbuilding. Not only would these untried, unproven fireboats be a first in the American fire service, assuming they performed as demanded, but this marked the first time a United States city went outside of the country to purchase fireboats. Historically, fireboats are most often built locally or regionally, especially if there exists longstanding shipbuilding design and construction expertise as has been true in the Pacific Northwest.

Parenthetically, we can only speculate over the additional derision accruing to the SES boats when the old-schoolers learned the boats would wear lime green color schemes. It is virtual holy writ in the fire service that no self-respecting fireboat or fire engine can be anything but red. Never mind that studies have shown that lime green enhances safety, because that color is more readily seen than red which is one of the most difficult colors in the spectrum to instantly recognize. Lime green or red: That controversial bit of trivia is likely to be debated among firemen well into the 21st Century as they grumble at the sight of lime green apparatus or fireboats.

The first SES boat to reach Tacoma was the *Defiance* which was lifted off the freighter *Hoegh Mallard* on May 25, 1982, and set down in Commencement Bay. The other SES craft, *Commencement*, was delivered several months later. After years of research, planning, design, controversy and skepticism, the moment of truth had arrived. Noted in passing was the fact that the boats cost more than they were budgeted, but Hovercraft assumed the loss as part of their investment in their confidence that SES boats would be ordered by other cities.

Tests showed the *Defiance* and the *Commencement* met or exceeded performance as well as design specifications. Tacoma fire officials and Vosper Hovercraft's designers and builders were more than vindicated. It was not long before disbelievers became believers and fireboat authorities came to Tacoma. They wanted to see how a moderately-sized fire department was able to achieve an answer to port protection problems which had vexed other cities, especially those troubled by initial costs, maintenance, and the cost-effectiveness of fireboats which no longer were mainly single-purpose vessels, but readily fulfilled a multiplicity of functions.

What they learned was not only a revelation, but dealt a significant blow to the idea that fireboats, at least as they have been known for over a century, are surely not anachronisms and doomed to ultimately vanish like dinosaurs. From performance and specifications viewpoints, as well as under emergency conditions, the SES boats are especially intriguing because they may well be a foretaste of future fireboat design and construction throughout the world.

Among the paramount features of the SES fireboats are the twin Detroit 8V92TI Marine diesel engines, each developing 445 horsepower at 2200 rpm, which are dedicated to propulsion. The lift engine also is a General Motors Detroit 6V92TI Marine diesel of 400 horsepower at 2300 rpm. The lift engine, when disengaged, powers one of the three Waterous single-stage centrifugal pumps nominally rated at 1833 gpm at 150 psi, but tests have proven they can do more than 2000. The two other Waterous pumps are driven by a Detroit 8V92TI Marine 570 horsepower diesel at 2200 rpm.

Powering each boat are three 24 volt, 125 amperes Leece-Neville alternators. Two of these are coupled to the propulsion engines and the third to the lift engine. Pumping tests prove the boats' capability of each delivering 7035 gpm at 100 psi. At 250 psi high pressure mode, the boats will discharge 4100 gpm. Access to the boats' machinery is by means of a deck hatch. There is no engine room.

All monitors are remotely-operated from the pilothouse, including the largest, a 5500 gpm Stang Hydronics master monitor mounted on the roof. Two Stang 2500 gpm monitors for underwharf or pier firefighting are mounted on the port and starboard sides of the bow. A pair of similar-sized monitors are located on the foredeck. The sixth gun, a Strebor, is attached to the high-level access tower ladder which, too, can be remotely operated from the pilothouse or from the base of the ladder. The telescoping ladder extends 35 feet and, with its hoist, can lift up to 2880 pounds of men and equipment onto decks of large vessels. Extended horizontally, the ladder can be used as an adjunct while making water rescues.

Additional firefighting capabilities are provided by the aft-mounted manually operated hydrants which can each supply one 5-inch water supply hose to land companies or two 2½-inch-diameter lines to firefighters. A third aft-mounted hydrant has outlets for two 1½-inch lines. That hydrant and the high-level access monitor are ideally equipped for dispersing around 1500 gpm of foam. Each boat carries about 150 gallons of fire extinguishing chemicals. More supplies can easily be provided by land companies.

How does Tacoma's visionary approach mesh with practicality? Ideally, according to Tacoma officials. Shortly after the *Defiance* went in service it sped to the Tacoma Narrows Bridge when a 19-year-old woman leaped into the water in an apparent suicide attempt. *Defiance* firefighters, arriving in less than 14 minutes, saved the teenager. The rescue was notable because it marked the first time in the 34-year his-tory of the bridge that a jumper had been saved. Previously, all 49 known jumpers drowned.

In September, 1984, the *Defiance* and the *Commencement* were called to battle a well-involved Point Defiance boat-house. Some 21 pleasure craft burned before the boats arrived, but the *Defiance* and *Commencement* were credited with delivering around 80 per cent of the water onto the fire and blocking it from spreading to exposed structures nearby.

Their multi-purpose capabilities were further demonstrated March 24, 1983, when the 574-foot-long *Ocean Steelhead*, a freighter loaded with logs, began taking on water at its berth and developed a serious list. After working for one day, salvage crews were not able to prevent the threat of imminent capsizing. The $35 million freighter with its $10 million cargo of logs was listing 41 degrees when the *Defiance* and the *Commencement* took stations at the freighter's bow and stern. With the help of land fire companies, the fireboats took turns dewatering the log-carrier. Inside of 30 minutes the list improved by 4 degrees. Around 12 hours later, the *Ocean Steelhead* was out of danger.

While major waterfront fires and capsizing vessels are infrequent occurrences, authorities are more apt to look at fireboats' multi-purposes and cost-effectiveness. Tacoma's SES boats can tow disabled boats weighing up to 97 tons, but a comparison between their fuel consumption and that of old *Fireboat No. 1* yields impressive figures.

"Tacoma's old gasoline-powered fireboat consumed 180 gallons of fuel an hour while operating at a flank speed of 12 knots, or 15 gallons per nautical mile," says Mitchell. "The SES fireboat uses approximately 2.5 gallons of diesel fuel per nautical mile at 30 knots speed, or one-sixth the amount of fuel used by the old fireboat. Significantly, the fireboat emergency response times have been improved by 60 per cent."

Not even the most staunch advocate of SES fireboats would say that they are the only answer to the ever-changing protection problems along the nation's waterfronts. A state of the art fireboat that meets the requirements of every port has never been built and probably never will. By 1989, Tacoma remained alone among American ports protected by SES fireboats. But it must be noted that other cities, notably New York, historically a harbinger of what's ahead in fire protection, has put a pair of SES fireboats high on its wish list.

The success of the SES boats led to the 1982 retirement of *Fireboat No. 1* after 53 years of service. In March, 1986, the waterfront landmark was taken from the water and put on permanent display in Tacoma's Marine Park on Ruston Way. The boat rests inside a dry berth resembling a graving dock that was designed by Naval Architect James Petrich.

Nowhere in the United States is there a more appropriate testament to the evolution of fireboats, a relatively unknown and unheralded slice of Americana than in Tacoma. As the venerable *Fireboat No. 1* lies in its final berth, the SES Fireboats *Defiance* and *Commencement*, stationed nearby, possibly portend the dawn of a new generation of fireboats. These three fireboats poignantly symbolize the dramatic history of fireboats, the incredibly spectacular fires that all fireboats fight and the courageous firemen who not only crew them, but sometimes give their lives while protecting American waterfronts.

Buried in smoke is a 200-foot-long Sinclair Refining Company warehouse containing 17,000 gallons of blazing petroleum products. Deeper in the smoke are 17 Sinclair oil storage tanks feeding the inferno in the Mill Basin section of Brooklyn. No fireboats were assigned when the first alarm was sounded at 3:46 p.m., May 10, 1962. As the blaze went to eight alarms, inaccessibility from the waterside resulted in the special call of three FDNY fireboats berthed about 18 miles away. First to arrive was the 8000 gpm *Gov. Alfred E. Smith* which opened with its deck pipes, including the bow gun which delivered 3500 gpm of foam. The fire was called the worst oil installation blaze in New York since 1919. From the Collection of the late Rev. Gordon W. Mattice

# Log of American Fireboats

This chronological log of 184 American fireboats with a pumping capacity of at least 1000 gallons-per-minute (GPM) merits footnoting. Most of these boats are more fully described in the text (see Index). Every effort was made to obtain at least two historically official verifications of the specifications of each boat. This task was complicated by conflicting information which might otherwise be considered definitive. Where this occurred, the most likely data was chosen for the log.

In several cases, GPM capacities could not be determined with exactitude, but other criteria established that these boats merited inclusion in the log.

Most of the discrepancies occurred in the lengths of the boats. Historians and maritime authorities are reminded that builders often cite two lengths: Overall length and length between perpendiculars. While it was not always possible to separate these figures, the overall lengths are most often cited.

GPM capacities also varied with individual boats. This is explained by differences in specifications and actual pumping tests. The latter, often as not, produce greater discharge readings. Because, moreover, fireboats invariably have longer service life spans than conventional fire engines, it was frequently found that GPM capabilities were increased during subsequent modifications following commissioning of the boats.

Historial records are vague or lost. This makes the task of compiling a definitive log fraught with peril. Extensive research provides strong evidence that a similar log has never been compiled.

No attempt was made to log boats delivering less than 1000 gpm. There were many of them, especially in recent years as fireboats trend toward smaller vessels. For the most part, however, these smaller boats primarily functioned—past and present—as patrol boats, auxiliary manpower and supply-carrying tenders, waterfront incident command posts, and, especially in contemporary times, for protection of small boat marinas, SCUBA-diver launching platforms, dewatering of boats in distress, search and rescue, and emergency medical or paramedic service providers.

| Year Built Or In Service | City | Name | Length | GPM Pumping Capacity |
|---|---|---|---|---|
| 1866 | New York | John Fuller | Unk. | Unk. |
| 1872 | Boston | William M. Flanders | 75' | 2500 |
| 1875 | New York | William F. Havemeyer | 106' | 2870 |
| 1875 | Philadelphia | William S. Stokley | 96' | 1000 |
| 1881 | New Orleans | Sampson | 108' | Unk. |
| 1881 | Philadelphia | Samuel G. King | 80' | 1200 |
| 1882 | New York | Zophar Mills | 125' | 2400 (6000?) |
| 1882 | Portland, Me. | Libby | Unk. | 6000 |
| 1885 | Chicago | Alpha | 77' | 3000 |
| 1885 | Chicago | Chicago | 90' | 2260 |
| 1886 | Brooklyn | Seth Low | 99' | 3360 |
| 1886 | Buffalo | George R. Potter | 80' | 3810 |
| 1886 | Chicago | W.H. Alley | 81' | 3000 |
| 1886 | Chicago | Geyser | 109' | 3810 |
| 1886 | Cleveland | Joseph L. Weatherly | 79' | 2400 |
| 1889 | Boston | John M. Brooks | 108' | 4240 |
| 1889 | Chicago | Yosemite | 109' | 5010 |
| 1889 | Milwaukee | Cataract | 106' 9" | 4660 |
| 1890 | New York | The New Yorker | 125' 6" | 11,530 |
| 1890 | Seattle | Snoqualmie | 90' | 5765 |
| 1891 | Baltimore | Cataract | 85' | 4400 |
| 1892 | Brooklyn | David A. Boody | 105' | 5765 |
| 1892 | Buffalo | John M. Hutchinson | 80' | 4445 |

| Year Built Or In Service | City | Name | Length | GPM Pumping Capacity |
|---|---|---|---|---|
| 1892 | Chicago | **Fire Queen** | 90' | 2260 |
| 1893 | Detroit | **Detroiter** | 115' | 5010 |
| 1893 | Milwaukee | **James Foley** | 99' | 5010 |
| 1893 | Philadelphia | **Edwin S. Stuart** | 110' 23" | 6000 |
| 1894 | Cleveland | **John H. Farley** | 90' | 3360 |
| 1894 | Cleveland | **Clevelander** | 82' | 5010 |
| 1895 | Boston | **Angus J. McDonald** | 110' | 5765 |
| 1897 | Erie, Pa. | **America** | Unk. | 2500 |
| 1897 | Erie, Pa. | **Erie** | Unk. | 2500 |
| 1897 | Milwaukee | **August F. Janssen** | 100' 6" | 4500 |
| 1898 | Chicago | **Illinois** | 118' | 10,000 |
| 1898 | New York | **William A. Strong (Robert A. Van Wyck)** | 110' | 6500 |
| 1900 | Buffalo | (renamed **Firefighter**, 1953; **Edward M. Cotter**, 1954) | 106' 3" | 9000 (later, 15,000) |
| 1900 | Detroit | **James Battle** | 116' | 7200 |
| 1902 | Detroit | **James R. Elliott** | 110' | Unk. |
| 1902 | Portland, Me. | **City of Portland (Engine 7)** | 87' 6" | 1650 |
| 1903 | Milwaukee | **MFD 15** | 106' 2" | 6000 |
| 1903 | New York | **Abram S. Hewitt** | 117' | 6000 |
| 1903 | New York | **George B. McClellan** | 117' | 7000 |
| 1904 | Portland, Ore. | **George H. Williams** | 105' 6" | 6200 |
| 1905 | New Orleans | **Sampson** (replacement) | 101' | 6000 |
| 1905 | Washington, D.C. | **Firefighter** | 101' 3" | 6000 |
| 1905 | Wilmington, N.C. | **Atlantic I** | 91' 3" | 6000 |
| 1906 | Milwaukee | **Cataract** | 88' | 4600 |
| 1908 | Duluth | **William M. McGonagle** | 125' | 12,000 |
| 1908 | Newark | **Newarker** | 63' | Unk. |
| 1908 | New York | **James Duane** | 132' | 9000 |
| 1908 | New York | **Cornelius Lawrence** | 105' | 7000 |
| 1908 | New York | **Thomas Willett** | 132' | 9000 |
| 1909 | Boston | **John P. Dowd** | 113' | 6000 |
| 1909 | Chicago | **Graeme Stewart** | 104' | 9000 |
| 1909 | Chicago | **Joseph Medill (I)** | 104' | 9000 |
| 1909 | Norfolk | **Vulcan** | 92' 4" | Unk. |
| 1909 | Seattle | **Duwamish** | 120' | 9000 (later, 22,800) |
| 1909 | San Francisco | **David Scannell** | 129' | 4000 |
| 1909 | San Francisco | **Dennis T. Sullivan** | 129' | 4000 |
| 1910 | Two Harbors, Minn. | **Torrent** | 110' | 12,000 |
| 1911 | Baltimore | **Deluge** | 120' | 9000 |
| 1911 | Boston | **Thomas A. Ring** | 84' 1" | 3000 |
| 1911 | Wilmington, N.C. | **Atlantic II** | 50' | 14,000 |
| 1912 | Portland, Ore. | **David Campbell** | 85' | 9200 |
| 1913 | Galveston | **Charles Clarke** | 100' | 4000 |
| 1915 | New York | **William J. Gaynor** | 118' | 7000 |
| 1916 | Cleveland | **Robert A. Wallace** | Unk. | 5000 |
| 1918 | Ketchikan | **Dale W. Hunt** | 33' 2" | 1000 (approx.) |
| 1918 | San Diego | **Bill Ketner** | 58' 9" | 5000 |
| 1919 | Los Angeles | **Fireboat No. 1 (Archibald J. Eley)** | 65' | 2000 |
| 1921 | Baltimore | **Cascade (S.C. 428)** | 110' | 2800 |
| 1921 | Baltimore | **Torrent** | 121' | 12,000 |
| 1921 | New York | **John Purroy Mitchell** | 132' | 9000 |
| 1921 | Philadelphia | **J. Hampton Moore** | 129' 3/4" | 10,000 |
| 1921 | Philadelphia | **Rudolph Blankenburg** | 129' 3/4" | 10,000 |
| 1922 | Jacksonville, Fla. | **John H. Callahan** | 110" | 7000 |
| 1922 | Milwaukee | **Torrent** | 110' | 12,000 |
| 1923 | New Orleans | **Deluge** | 139' 7" | 10,000 |
| 1925 | Houston | **Port Houston** | 118' | 10,000 (approx.) |
| 1925 | Los Angeles | **Los Angeles No. 2 (Ralph J. Scott)** | 99' | 13,500 (later, 18,655) |
| 1927 | Portland, Ore. | **David Campbell** | 87' 6" | 12,500 X |

| Year Built Or In Service | City | Name | Length | GPM Pumping Capacity |
|---|---|---|---|---|
| 1927 | Portland, Ore. | **Mike Laudenklos** | 87' 6" | 12,500 |
| 1927 | Seattle | **Alki** | 123' 6" | 12,000 |
| 1928 | Portland, Ore. | **Karl Gunster** | 87' 6" | 12,000 |
| 1929 | Detroit | **John Kendall** | 128' 3" | 12,000 |
| 1929 | Galveston | **City of Galveston** | 82' 5" | Unk. |
| 1929 | Tacoma | **Fireboat No. 1** | 96' 6" | 10,000 |
| 1931 | Boston | **Matthew J. Boyle** | 125' | 12,000 |
| 1931 | Portland, Me. | **City of Portland** (probably a replacement) | 90' | 6000 |
| 1931 | New York | **John J. Harvey** | 130' | 16,000 |
| 1932 | Wilmington, N.C. | **Atlantic IV** | 64' | 2500 |
| 1937 | Chicago | **Fred A. Busse** | 90' 6" | 10,000 |
| 1938 | New York | **Fire Fighter** | 134' | 20,000 |
| 1939 | Mobile | **Ramona Doyle** | 63' | 3800 |
| 1940 | Pearl Harbor | **Hoga** (later, **Port of Oakland, City of Oakland**) | 99' 7" | 4000 (later, 10,000) |
| 1942 | New Orleans | **Bourgeois II** | 94' | 6000 |
| 1942 | New Orleans | **John McGiven** | 102' 6" | 10,000 |
| 1945(circa) | Beaumont, Tex. | - - - - - - - - - | 65' | 1000 |
| 1945 | Ketchikan | **H.V. Newell** | 65' | 4000 |
| 1946 | Cleveland | **Marvet** | 40' | 2000 |
| 1946 | Newark | **William J. Brennan** | 40' | 3200 |
| 1946 | Newark | **Michael P. Duffy** | 40' | 3200 |
| 1946 | Washington, D.C. | **William T. Belt** | 112' | 4000 |
| 1947 | Boston | **James F. McTighe** | 97' | 6000 |
| 1947 | Boston | **Joseph J. Luna** | 97' | 6000 |
| 1947 | San Francisco | **Frank G. White** | 72' | 6000 |
| 1948 | Philadelphia | **Bernard Samuel** | 75' 10" | 6000 |
| 1948 | Wilmington, N.C. | **Atlantic III** | 65' | 2000 |
| 1949 | Chicago | **Joseph Medill (II)** | 92' | 12,000 |
| 1949 | Chicago | **Victor L. Schlaeger** | 92' | 12,000 |
| 1949 | Milwaukee | **Deluge** | 97' 6" | 12,000 |
| 1950 | Houston | **Captain Crotty** | 79' 4" | 6000 |
| 1950 | Philadelphia | **Benjamin Franklin** | 79' 4" | 6000 |
| 1950 | Philadelphia | **Delaware** | 79' 4" | 6000 |
| 1951 | Honolulu | **Abner T. Longley** | 87' | 9000 |
| 1951 | Jacksonville, Fla. | **Richard D. Sutton** | 104' | 10,000 |
| 1953 | Buffalo | **Engine 29 ("Duck")** | Unk. | 1000 |
| 1953 | Long Beach, Ca. | **Fireboat 15** | 58' 5½" | 4500 |
| 1954 | Cleveland | **Clevelander** | 58' 4" | 6000 |
| 1954 | Ketchikan | **Harry V. Newell** | 65' | 4000 |
| 1954 | Long Beach, Ca. | **Fireboat 20** | 58' 5½" | 4500 |
| 1954 | New York | **John D. McKean** | 129' 9" | 19,000 |
| 1954 | San Francisco | **Phoenix** | 88' | 9600 |
| 1956 | Baltimore | **Mayor Thomas D' Alesandro, Jr.** | 104' | 12,000 |
| 1956 | Pittsburgh | **Cornelius D. Scully** | 70' | 10,000 |
| 1956 | Tampa | **Francis Bellamy** | 68' 2" | 6000 |
| 1958 | Eureka, Ca. | **Capt. Pudgy Davis** | 40' | 2400 |
| 1958 | New York | **Dr. Harry M. Archer** | 105½' | 8000 |
| 1958 | New York | **Sen. Robert F. Wagner** | 105½' | 8000 |
| 1958 | New York | **Smoke II (Tender)** | 51' 9" | 2000 |
| 1958 | New York | **Sylvia 'H.G.' Wilks** | 105½' | 8000 |
| 1958 | Stockton, Ca. | **Mistress Delta** | 40' | 2000 |
| 1959 | Portland, Me. | **City of Portland** | 64' 10½" | 7000 |
| 1960 | Baltimore | **August Emrich** | 85' | 6000 |
| 1960 | Baltimore | **Mayor J. Harold Grady** | 85' | 6000 |
| 1960 | Baltimore | **P.W. Wilkinson** | 85' | 6000 |
| 1960 | Fort Lauderdale, Fla. | **Fireboat 1 (Old Reliable)** | 28' | 1500 |
| 1961 | Cleveland | **Anthony J. Celebreeze** | 60' | 6000 |
| 1961 | New York | **Gov. Alfred E. Smith** | 105' 6" | 8000 |
| 1961 | New York | **John H. Glenn, Jr.** | 70' | 7000 |
| 1961 | San Diego | **Louis Almgren** | 65' | 4000 |
| 1961 | Tampa | **Gail Holland** | 64½' | 2000 |
| XI 1962 | Los Angeles | **Fireboat No. 4** (later **Bethel F. Gifford**) | 76½' | 9000 |

| Year Built Or In Service | City | Name | Length | GPM Pumping Capacity |
|---|---|---|---|---|
| 1962 | New Haven | Sally Lee | 69' 6" | 7000 |
| 1963 | Jacksonville, Fla. | Fireboat 82 | 63' | 4000 |
| 1963 | New York | Flame | 17' 10" | 1100 |
| 1964 | Newark | John F. Kennedy | 46' | 4000 |
| 1968 | Cincinnati | Dr. Giles DeCourcy | 26' | 5000 |
| 1969 | Fort Lauderdale | Striker | 37' | 1250 |
| 1969 | Jacksonville, Fla. | Eugene Johnson | 65' | 6000 |
| 1970 | New York | James F. Hackett | 29' 3" | 2500 |
| 1971 | San Diego | Harbor Island | 42' | 4500 |
| 1971 | Wilmington, N.C. | Atlantic V | 61' | 4000 |
| 1972 | Boston | Firefighter | 76' | 6000 |
| 1972 | Boston | Howard W. Fitzpatrick | 76' | 6000 |
| 1972 | Portland, Ore. | Spencer | 43' | 6000 |
| 1972 | San Diego | Point Zuniga | 32' | 1200 |
| 1974 | Bristol, Pa | - - - - - - - | 16' | 3125 |
| 1974 | Houston | Capt. Farnsworth | 80" | 6000 |
| 1974 | Tampa | Francis Bellamy | 64' | 6000 |
| 1975 | Mobile | Ramona Doyle | 65' | 3000 |
| 1976 | Boston | St. Florian | 45' | 3000 |
| 1976 | Los Angeles County | Fireboat 110 | 37' | 1500 |
| 1977 | Detroit | Curtis Randolph | 77' 10" | 10,000 |
| 1977 | San Diego | Point Loma | 32' | 1500 |
| 1978 | San Diego | Shelter Island | 32' | 2000 |
| 1979 | Miami | City of Miami | 35' | 2500 |
| 1980 | Cincinnati | David T. Sheehy | 25' | 1000 |
| 1980 | Port Everglades, Fl. | Dusky | 23' | 12500 |
| 1982 | Tacoma | Commencement | 70' | 7000 |
| 1982 | Tacoma | Defiance | 70' | 7000 |
| 1983 | Houston | J.S. Bracewell | 68' | 4000 |
| 1983 | Houston | H.T. Tellepsen | 68' | 4000 |
| 1983 | Long Beach, Ca. | Fireboat 21 | 35' | 1000 |
| 1983 | Portland, Ore. | George H. Williams (replacement) | 40' | 2500 |
| 1984 | Milwaukee | Amphibian | 35' | 2500 |
| 1984 | Seattle | Chief Seattle | 96' | 8500 |
| 1984 | Wilmington, N.C. | Fort Johnston | 106' | 2000 |
| 1986 | Ketchikan | Harry Newell (replacement) | 45' | 5000 |
| 1986 | Long Beach, Ca. | Challenger | 88' 6" | 10,000 |
| 1986 | Stockton, Ca. | DeKaury | 100' | 4000 |
| 1987 | Des Moines, Wa. | Fire-Rescue | 24' | 1250 |
| 1987 | Long Beach Ca. | Fire-Rescue I | 24' | 1250 |
| 1989 | Port Everglades, Fl. | Captiva | 24' | 1200 |
| 1989 | Honolulu | (Under construction, not named) | 110' | 7000 |

# ACKNOWLEDGEMENTS

One of the pleasures of writing is the new friends you make and the renewed association with old ones. Although I have been collecting archival and other fireboat information for many years and have been aboard fireboats in battle, the book would not have been possible without the help, encouragement and access to research and illustrative material that was so freely given by those who took an active interest and shared my enthusiasm.

Engrossed as he was with research and development of Los Angeles' proposed state of the art fireboat, Pilot Bill Dahlquist of the LAFD's **Ralph J. Scott** and a fellow historian, made time to read portions of the book, to make suggestions and to patiently answer dozens of questions that arose during the writing of **Fireboats.** No wonder he was named 1982 Los Angeles Fireman of the Year.

No less an active interest and encouragement came from:

James P. Delgado, Maritime Historian and Head of the National Maritime Initiative, National Park Service. Not only did he write the Foreword, for which a special note of gratitude is due, but wherever his travels took him, he helped in countless ways and opened doors which might otherwise be closed to me. My active interest and participation in his National Historic Landmarks project pertaining to fireboats was both an honor and a long-felt wish to see American fireboats recognized for their role in history.

John Waterhouse, President, Elliott Bay Design Group, Ltd., Seattle, who, as an internationally-recognized fireboat designer and authority, read the entire manuscript and made many suggestions for inclusion in the final manuscript.

Bob Freeman, historian and retired member of the Chicago Fire Department, gave generously of his time, his vast photo collection and otherwise encouraged me, both as a friend and colleague, when my desire was there but my two typing fingers faltered.

Jim Murray, retired Marine Engineer of the Fire Department of New York's Fireboat **Fire Fighter,** and historian as well, not only read technical portions of the manuscript regarding that world's most famous fireboat for clarity and accuracy, but counseled on the history of the legendary FDNY Marine Division, while he patiently researched and verified my photo information.

Steven Lang, co-author with Peter H. Spectre, of **On The Hawser, A Tugboat Album,** who generously made his unpublished fireboat photo collection available to me and otherwise encouraged this project. **On the Hawser** complements **Fireboats** because the two craft are so similar, if not in function, surely as unheralded waterfront landmarks. Readers of **Fireboat** will, no doubt, want to read their book published by Down East Books, Camden, Me., 1980.

Sadly, two long-time colleagues will not see the results of their assistance for which I will forever be profoundly grateful:

The late Robert Burns, retired Battalion Chief of the Philadelphia Fire Department, whose pictures, research and suggestions played a vital role in the Philadelphia sections of the book. Bob busied himself with research for me right up to the day of his death.

The late Clarence E Meek, retired Assistant Chief of the Fire Department of New York. Clarence, a friend of many years when we served as Contributing Editors of Fire Engineering, pioneered in researching and writing the history of FDNY and other fireboats. Clarence, many years ago, encouraged me to undertake a complete history of the American fireboat and I hope he would have been pleased with the results.

Significant contributions, including research, photos, illustrations, fact-checking and reading portions of the book while making suggestions for accuracy were made by:

Keith Allen, Peter E. Balducci, Walter Ball, Dick Bangs, George "Smokey" Bass, Ernest Belles, Kenneth Besman, Robert E. Brewis, Robert Brown, Buffalo Fire Historical Society, Bill Burmester, Jim Choner, Eric W. Caplan, Gary Carino, Paul Christian, Michael J. Cline, Vern Cooney, Gary Conklin, Tony DiDomenico, Charles Cornell, Francis J. Duffy, Albert Duke, W.C.Dunn, Bettye Ellison, Herbert Eysser, Kevin Foster, Gary E. Frederick, Louis Galante, Glendale, Ca., Public Library Computerized Reference Desk, William J. "Bill" Guido, Claude Harris, Richard Heath, Stephen G. Heaver, Jr., Dennis Hedrick, David C. Henley, Edward C. Henry, Ray Henry, Don Hibbard, Sam Hill, Harry Hotston, Roy J. Johnson, Debbie Kraybill, Jerry Lawrence, Warner L. Lawrence, Jack Lerch, Dan Martinez, Vincent L. Marzo, the late Rev. Gordon W. Mattice, Mike Meadows, William A. Miller, Patty Milner, Tony Mitchell, Harry Morck, Phil McBride, Joseph B. McManus, Maynard McQuaw, Dick Neal, William "Bill" Noonan, Donald Parrot, Jim Perry, Paul Perry, Gary Pirkig, Allen W. Radcliffe, Gerald Ramaekers, Roger Ramsey, John Rassmussen, Bruce Reagan, the late Albert Redles, Ed Reed, Jack Robrecht, David S. Rogers, Richard Roper, Steven Scher, Mort Schuman, Kenny R. and Richard J. Sikora, the Society of Naval Architects and Marine Engineers, Southern California Answering network (SCAN) of the Los Angeles Public Library, Jack Supple, Gary A. Svider, Ronald J. Sylvia, Stanley L. Thaut, Don Thomas, George Thornhill, Kenneth Trometer, John C. Ware, Fred Warner and the Washington State Historical Society, Tacoma.

I would, moreover, be remiss without acknowledging the encouragement of my publisher, Fred Conway, President of Conway Enterprises, Inc., who held a steady course when there were some of us who were about to become scuppersawash in those inevitable gliches which occur in all books from conception to delivery. My thanks, too, to the staff and associates of Conway Enterprises, Inc., who enthusiastically assisted with this book in so many ways.

Paul Ditzel
Woodland Hills, Ca.

# BIBLIOGRAPHY

**A Synoptical History of the Chicago Fire Department.** Chicago: Benevolent Association of the Paid Fire Department of Chicago, 1908.

"Amphibious fire boat for Miami." **Marine Engineering/Log**, April, 1983; p. 45.

Baird, Donal M. "Fireboats: A Century of Development." **Fire Engineering**, Aug. 1966; p. 136.

Baker, Thomas E. "New York City's New Fireboats." **Marine Engineering/Log**. July, 1958; p. 59.

"Baltimore City FD Gets New 24 Ft. Fire/Rescue Boat." **Pennsylvania Fireman**, Feb. 1989; p. 88.

"Baltimore's Fleet of Fireboats." **Firemen**, Nov. 1961; p. 12.

Casey, James F. "Brooklyn Mill Basin Fire, Largest oil blaze in New York since 1919." **Fire Engineering**, Aug. 1962; p. 634.

Churchill, Winston S. **The Second World War**. Vol. II, "Their Finest Hour". Boston: Houghton Mifflin Co., 1949.

Clevenger, Mark T. "New 'insurance' for Seattle Waterfront." **The Work Boat**, March, 1985; p. 45.

_____ . "Washington—Something Old, Something New." **The Pacific Maritime Magazine**, Dec. 1984; p. 24.

Coakwell, Frank, C.E. "Toronto Fights Ship Fire." **Fire Engineering**, Jan. 1966; p. 30.

_____ . "Toronto's Fireboat." **Firemen**, June, 1967; p. 37.

Coleman, H.C. "Much Marine Electrical Progress During 1938." **The Marine News**, March, 1939; p. 25.

Cudahy, Bryan J. "The Fireboats of New York." **Sea Classics**, Nov. 1974; p. 18.

Dahlquist, William E. "Port and Fireboat Survey 1986." Los Angeles Fire Department, 1986.

_____ . "New Power For The Scott." **The Firemen's Grapevine**, Aug. 1978; p. 8.

_____ . "Fire On The Waterfront, A History of Fire Protection in Los Angeles Harbor (1542-1984)." **The Firemen's Grapevine**, July, 1984; p. 24.

Daly, Joseph P. "5-Alarm Fire Destroys Pier, Perils Ship in San Francisco." **Fire Engineering**, July, 1973; p. 36.

**Dictionary of American Naval Fighting Ships**. Navy Department, Office of The Chief of Naval Operations, Naval History Division. Vol. III, 1968; p. 342.

Ditzel, Paul. **A Century of Service, 1886-1986, The Centennial History of the Los Angeles Fire Department**. New Albany, Ind; Conway Enterprises, Inc., 1986.

_____ . "Cast Off and Pray! Los Angeles **Fireboat 2** Raced Into Action as Explosions Rocked the Waterfront." **Man's Magazine**, Vol. 8, No. 5; May, 1960; p. 16.

_____ . "Fireboats... 'Dispatch All Boats!'." **The Firemen's Grapevine**, March, 1989; p. 15.

_____ . "Endangered Species? Changing needs launch new wave of fireboats." **Firehouse**, Aug. 1981; p. 60.

_____ . **Fire Engines Firefighters**. New York: Crown Publishers, Inc., 1976.

_____ . **Firefighting, A New Look in the Old Firehouse**. New York: Van Nostrand Reinhold Co., 1969.

_____ . "Four Chicago Firemen Killed at 5-11 Waterfront Fire." **Fire Engineering**. Feb. 1951; p. 104.

_____ . "Los Angeles Fights Fire with Scuba." **Fire Engineering**, Nov. 1965; p. 36.

_____ . "Los Angeles Firemen Avert Potential Harbor Disaster." **Fire Engineering**, Aug. 1954; p. 652.

_____ . "On the Job—Los Angeles. LAFD Fireboats Avert Potential Wharf Disaster." **Firehouse**, Jan. 1989; p. 59.

_____ . Tapes and transcripts of interviews with Robert Brown, Quartermaster, YT-146 (**Hoga**) during the Dec. 7, 1941, Pearl Harbor attack: 11/20/88, 11/21/88.

_____ . Tapes and transcripts of interviews with Joe B. McManus, Chief Boatswain's Mate, YT-146 (**Hoga**) during the Dec. 7, 1941, Pearl Harbor attack: 11/22/88.

_____ , with Gray, Brainard. "Even the Water Burned." **Male Magazine**, Vol. 4, No. 6; June, 1954; p. 11.

_____ . "World's Mightiest Fire Engine." **Popular Science**, Vol. 191, No. 4; Oct. 1967; p. 96.

Douglas, J.S. "Berth 174 Wharf Fire." **The Firemen's Grapevine**. March, 1968; p. 4.

Duffy, Francis J. "1904; Slocum in Flames." **Firehouse**, Aug. 1979; p. 30.

Elie, Curt. "The Pride of the Fleet." (Portland, Me.) **Fire Apparatus Journal**, Jan. 1988; p. 12.

"'Empty' Ship Blast Kills 5; 60 Hurt." **Western Fire Journal**, Feb. 1977; p. 8.

"Fast-Response Fireboat for Seattle." **Harbour & Shipping**, April, 1985; p. 29.

"Fire Boat **Deluge**." **Marine Engineering and Shipping Review**, Dec. 1949; p. 42.

"Fire Boats **David Scannel** And **Dennis T. Sullivan** for San Francisco Harbor." **International Marine Engineering**, June, 1910; p. 244.

**Fire Boats**. General Electric Company, Schenectady, N.Y., Power and Mining Department. Bulletin No. 4702; Oct. 1909.

"Fireboat **Chief Seattle** To Replace **Duwamish**." **Marine Digest**, Oct. 20, 1984; p. 11.

"Fireboat 5" (Tacoma). **The Work Boat**, July, 1983; p. 35.

"Fireboat takes DC by storm." **Fire Command**, Dec. 1977; p. 17.

"Fireboats, A Survey of These Special Units in Major Cities." Part 1, **Firemen**, Feb. 1966; p. 21; Part 2 **Firemen**, March 1966; p. 15.

"**Fire Fighter**, Most Powerful Fireboat." **Motorship and Diesel Boating**, Nov. 1938; p. 20.

"Fire Rescue Boats." **News Letter, Monthly Publication of The Los Angeles Fire Department**. Jan. 1967; p. 1.

"First S.E.S. Multi-Purpose Fireboat In The World." City of Tacoma, Wash. Undated pamphlet, courtesy Tacoma Fire Department.

Fitzsimmons, Matthew. "The Night the Sea Caught Fire." **Guideposts**, June, 1976; p. 56.

Frederick, Gary. "Shipshape in Baltimore, A history of the city's fireboat fleet." **Firehouse**, June 1986; p. 49.

Freeman, Robert. "Chicago Fire Boats, 1870-1949". Privately published, c.1949.

_____ . "Fire Department New York: Fireboats & Marine Companies, Chronology, Origin, Nativity". Privately published. 1983.

_____ . "The Fleet Is in." Part 1, **The Chicago Fire Fighter**, Winter, 1968; p. 9; Part 2, **The Chicago Fire Fighter**, Spring, 1968; p. 33.

Gardner, Rich; Robrecht, Jack. "A History Of The Philadelphia Fire Boats." **Firemate**, Sept., 1979; p. 7.

Green, Mike. **Massey Shaw— The Man and the Boat**. Pamphlet, courtesy of the Massey Shaw and Marine Vessels Preservation Society, Ltd., Honorable Secretary David S. Rogers, Romford, Eng.

Greenock, Bob. "Ammunition Blast Demolishes Ships and Shore Structures." **Fire Engineering**, Aug. 1944; p. 543.

Hageman, Edward C. and Gow, Paul A. "**Chief Seattle**, A Case Study of the Design Process." Nickum & Spaulding Associates, Inc., Seattle, with the cooperation of the Seattle Fire Department. Presented to the International Work Boat Show, New Orleans, Jan. 1985.

"**Harry Newell**." **The Work Boat**, Nov. 1986; p. 19.

Hecker, Klaus P. **International Fireboats**. Hanover, Germany: EFB-Verlagsgesellschaft mbH., 1982.

Henley, David C., Brig. Gen., Nevada National Guard (Reserve). **Battleship Nevada, The Epic Story of the Ship that Wouldn't Sink**. Fallon, Nev.: Western America History Series, Lahontan Valley Printing, Inc., 1988.

Johnson, Gus. **F.D.N.Y., The Fire Buff's Handbook of The New York Fire Department, 1900-1975**. Belmont, MA: Western Islands, 1977.

Johnston, W.W., Jr. "Tacoma, Washington Dock Fire." **The Firemen's Grapevine**, Sept. 1963; p. 10.

_____ , and Egerton, Cal. "Scuba Fire Fighting on the Barbary Coast." **The Firemen's Grapevine**, Feb. 1967; p. 4.

King, Frank L. **The Missabe Road... The Duluth, Missabe and Iron Range Railway**. San Marino, CA: Golden West Books, 1972.

Kozak, Daniel. "Heroism at Sea, Firefighters vs. Hazardous Materials on the **S.S. Sea Witch**." **Firehouse**, Dec. 1979; p. 44.

Kull, Irving S. and Nell M. **A Chronological Encyclopedia of American History**. New York: Popular Library, 1969.

Lawrence, Warner L. "Modernized Fireboat." **The Firemen's Grapevine**, Feb. 1970; p. 7.

Lord, Walter. **Day of Infamy, The Moment-By-Moment Story of Pearl Harbor, December 7, 1941**. New York: Bantam Books, Holt, Rinehart & Winston, Inc., 1957.

Lott, Arnold S., USN (Ret.) and Sumrall, Robert F. HTC, USNR. **Pearl Harbor Attack, (An Abbreviated History)**. Annapolis, Md.: 1977.

"Mammoth Fire Boat Designed by H.S. Maxim, M.E." **Scientific American**, Vol. XLIV, No. 10, March 5, 1881; p. 1.

**Marine Board Casualty Report SS Sansinena (Liberian); Explosion and Fire in Los Angeles Harbor, California on 17 December 1976 with loss of life**. U.S. Coast Guard, Washington, D.C., Report No. USCG 16732/71895. Nov. 25, 1977.

"Marine Companies, Fire Department-City of New York." **W.N.Y.F.**, 1963. p. 15.

Maslin, Marsh. "City's New Fireboat Will Be 'The Phoenix'." **San Francisco Call-Bulletin**, Jan. 7, 1954.

Meek, Clarence. "F.D.N.Y. Flotilla, A History of our Fireboat Fleet Through the Years". **W.N.Y.F.**, First Issue, 1969; p. 16.

_____ , "Fireboats Through The Years." **W.N.Y.F.**, July, 1954, p. 24.

_____ , "Log of the Nation's Fireboat Fleet." Part 1, **Fire Engineering**, Aug. 1957, p. 758; Part 2, **Fire Engineering**, Sept. 1957, p. 938.

Milnes, Robert B. "Firefighting Amphibian." **Firehouse**, Jan. 1983; p. 43.

Mitchell, Tony F. "A Ten-Year Dream Fulfilled: Tacoma's New Fireboats." **Proceedings of Marine Safety Council, U.S. Coast Guard**, April, 1985; p. 87.

"Tacoma's Boats Speed In New Era Of Marine Fightfighting." **American Fire Journal**, Oct. 1983; p. 45.

"Motor Fire-Boats—A West-Coast Development." **Standard Oil Bulletin**, April, 1928; p. 3.

"Multi-Purpose Surface Effect Ship Fireboat." Pamphlet. City of Tacoma Fire Department. Undated.

Murray, Jim and Trojanowicz, Al. "The **Fire Fighter**—Protecting New York Harbor for 50 Years." **Fire Apparatus Journal**, July, 1988; p. 8.

"N.Y.C. Gets New Fire Fighter." **Marine Engineering**, Sept. 1954; p. 54.

Nailen, R.L., and Haight, James S. **Beertown Blazes, A Century of Milwaukee Firefighting**. Milwaukee: NAPCO Graphic Arts, Inc., 1971.

"Newark's New Fireboat." **Fire Command**, Jan. 1965; p. 10.

"New York Fireboats". **Tow Line**, Moran Towing & Transportation Co., Inc., New York, Spring, 1970.

Noonan, William F. **Smoke Showin'**. Boston: Addison C. Gethcell & Sons, Inc., 1984.

McCullough, K.E. "Space-Age Pump Installed on Amphib Gives Miami a 2500-GPM Fireboat." **Fire Engineering**, Dec. 1979; p. 20.

McManus, J.B. CBM., USN, Tugmaster, YT-146. **Log of YT-146, Dec. 7, 1941**. Pearl Harbor, Hawaii: United States Navy Yard. Yard Craft Office, 1941.

McNairn, Jack. "Fire Boats On San Francisco Bay." **Pacific Coast Fire Journal**, Sept. 1958; p. 12.

McVann, William F. "Grace Line Pier Fire." **W.N.Y.F.**, Jan. 1948; p. 14.

"1 Dead, 28 Burned, 3 Boats Lost As Oil Fire Sweeps Buffalo River." **Buffalo Evening News**, July 27, 1928; p. 1.

"Queen of the fire fleet, Baltimore's fireboat **Mayor Thomas D' Alesandro, Jr.**" **Fairbanks-Morse News**, Jan.-Feb., 1957; p. 4.

Parsons, Harry deBerkeley. "**American Fireboats**." Paper read at the fourth general meeting of the Society of Naval Architects and Marine Engineers, New York Nov. 12-13, 1896. Transactions of the Society of Naval Architects and Marine Engineers, Vol. IV, 1896.

"Philadelphia Fireboat **Bernard Samuel**." **Marine Engineering and Shipping Review**, Jan. 1949; p. 72.

"Pier #7, Port of Tacoma... Fire Investigation, Fatality." Fire Department, City of Tacoma, July 14, 1963.

Pirkig, Gary. "Fighting fire while surrounded by water: Firefighters train on floating highrise." **The California Fireman**, Feb. 1988; p. 18.

Prange, Gordon W., in Collaboration With Goldstein, Donald M. and Dillon, Katherine V. **At Dawn We Slept, The Untold Story of Pearl Harbor**. New York: Viking Penguin, 1981.

_____ . **Dec. 7, 1941, The Day The Japanese Attacked Pearl Harbor**. New York: McGraw-Hill Book Co., 1988.

_____ . **Pearl Harbor, The Verdict of History**. New York: McGraw-Hill Book Co., 1986.

"Protection of Water Fronts." **Fireman's Herald,** Vol. 66, No. 25, Dec. 19, 1914; p.1.

"Proto." **The Massey Shaw At Dunkirk, The Association of Dunkirk Little Ships.** Monograph, courtesy of the Massey Shaw and Marine Vessels Preservation Society, Ltd., Honorable Secretary David S. Rogers, Romford, Eng.

"**Report of Conflagration In Grain Elevators Operated By Rosenbaum Bros. Inc. and Norris Grain Company, Chicago, Illinois, May 11, 1939.** Published By The Chicago Board of Underwriters, 1939.

Rice, William T., Capt., USNR (Ret.). **Pearl Harbor Story, Authentic information and pictures of the attack on Pearl Harbor, December 7, 1941.** Honolulu: Swak, Inc., 1965.

Riepe, Bill. "Inferno in the Harbor, Munitions ship goes ablaze on New Jersey Waterfront during WWII." **Firehouse**, April 1988; p. 63.

Robinson, Wayne. "Fire on the Water. Philadelphia's fireboats: They're old, but they get the job done." **The Sunday Bulletin/Discover**, Aug. 14, 1977; p. 13.

"San Diego Seems Satisfied." **The Pacific Maritime Magazine,** Dec. 1984; p. 31.

"San Francisco's Fireboat **Phoenix**." **Marine Engineering,** Aug. 1954; p. 48.

Scott, C.S. Stone. **Pearl Harbor, The Way It Was — December 7, 1941.** Honolulu: Island Heritage Limited, 1977.

"Sea Witch — Esso Brussels." Extract from the Marine Casualty Report (No, USCG/NTSB-MAR-75-6) released March 2, 1976. **Proceedings of the Marine Safety Council of The United States Coast Guard,** May, 1976; p. 86.

Snyder, William F., and Murray, William A. **The Rigs Of The Unheralded Heroes.** Baltimore: E. John Schmitz & Sons, Inc., 1971.

Stevens, A.D. **"A 1650 Horse-Power Gasoline Fire Boat."** Read at the 30th general meeting of the Society of Naval Architects and Marine Engineers, New York, Nov. 8-9, 1922.

Supple, Jack. **History Of The Buffalo Fire Department, 1880-1979.** Buffalo: Privately-published, 1980.

"Surplus fireboat again covers the waterfront." (Mobile, Ala.) **Fire Command,** Dec. 1975; p. 19.

"Tacoma's New Fireboat." **Railway and Marine News,** July, 1929; p. 25.

"Tacoma Tries New Concept In Marine Firefighting." **Western Fire Journal,** Dec. 1981; p. 35.

"Tacoma's new fireboats fly to the rescue as well." **Nor'westing,** June, 1983; p. 12.

Thau, W. E. **"Diesel Electric Propulsion."** Read at the 34th general meeting of the Society of Naval Architects and Marine Engineers, New York, Nov. 11-12, 1926.

"The Burning Of The North German Lloyd Steamships And Docks, Hoboken, N.J." **Scientific American.** Vol. XLVI, July 14, 1900; p. 646.

"The Fire Fleet, 'Watchdogs of the Waterfront." **W.N.Y.F.,** July, 1944; p. 18.

The World At War. Vol. 6. Thames Television. 1980. Thorn EMI Video.

**"The Grand Trunk Pacific Dock Fire of July 30, 1914.** From the 24th Annual Report of the Fire Department of The City of Seattle, 1914.

**The La France Fire Engine Company.** Elmira, NY: Catalog, 1898.

"30-knot fireboat for Ketchikan." **The Pacific Maritime Magazine,** Aug. 1986; p. 9.

"Veteran Seattle Fireboat To Be Retired Soon." **Marine Digest,** March 17, 1984; p. 11.

Walsh, Thomas P.J. "The **Sea Witch**...and a cauldron of fire and death." **W.N.Y.F.,** 4th Issue, 1973; p. 4.

Ward, J.T., and Casey, J.F. "Aircraft Carrier **Constellation** Fire." **Quarterly of the National Fire Protection Association,** April, 1961; p. 283.

Werner, William. **History Of The Boston Fire Department and Boston Fire Alarm System, January 1, 1859 Through December 31, 1973.** Boston: The Boston Sparks Association, 1973.

Werner, William; Kelleher, Richard; Noonan, William F.; and Fitzgerald, Grank W. **"Fireboats of The Boston Fire Department."** Boston Sparks Association: 1984.

West, Charles C. **"Centrifugal Pump Fire-Boats."** Read at the 16th general meeting of the Society of Naval Architects and Marine Engineers, New York, November 19-20, 1908.

Wilhelm, Steve. "Fire Power." **Marine Digest,** April 23, 1988; p. 11.

"Tacoma's Fireboat Saved For the Future." **Marine Digest,** May 10, 1986; p. 11.

Williams, Harold A. **Baltimore Afire,** Revised Edition. Baltimore: Schneidereith & Sons, 1979.

"Workboats Northwest builds versatile fire/rescue boat." **The Work Boat,** Oct. 1987; p. not shown.

"World's Most Powerful Fireboat, Diesel-Electric Fire Fighter for New York City to Have Unusually Elaborate Equipment." **Motorship and Diesel Boating,** Dec. 1937; p. 16.

Zegarelli, Mark. "The Hoboken Horror, 361 Die in Pier Disaster." **Firehouse,** May, 1979; p. 44.

Zini, Frank. "American Fireboats." **Firehouse**, May, 1978; p. 26.

Zitko, Lee and Wigfield, Mike. "The Kid Sisters." **The Firemen's Grapevine,** Sept. 1967; p. 5.

# INDEX

"A Synoptical History of the Chicago Fire Department", 34

*Admiral Farragut,* 40

Admirals
Meade, Richard W., 17
Nimitz, Chester W., 78
Ramsey, Sir Bertram, 82

Admiralty Board, 82

*Aeolian,* 45, *45*

Ahrens-Fox pump, 63

Aiea Bay Mudflats, 75

Aircraft Carrier, *Constellation,* 71

Aircraft
Army, Marine, Navy, 76

Alden, John G., 94, *95,* 97, 98, *100*

Alderson, John H., 85, 91

Alexander Miller and Brothers, shipbuilders, 29

Algren, Louis, 43

*Alki,* 56, *56,* 58, 59, 115, *115,* 116

Allen, Maurice P., 85

Allied Warships, 80

*America,* 64

"America Corn-Pith Cellulose", 21

American Export Isbrandtsen ship
*Sea Witch, 108,* 109, *109*

"American Fire-Boats", 9

American Fire Engine Co., *11*

American Merchant Marine Seamanship Trophy, 71
See Awards

American President Lines
Wharf and warehouses, *90,* 90, 91

Amoskeag Manufacturing Works, 9, *20,* 21, *27*

*Amphibian,* 115

*Anna Marie,* 4, 8

Apparatus Operator
Zar, Tony, *7,* 8

Armour's Union Elevator, 34

Astronaut
Glenn, John H., Jr., *100*

*Athlon,* 40

Atlantic Ocean, 91

Atlas Oil Company, 58

Auxiliary Fire Service
Jackson, Frank Whitford, 82
Ray, Henry Albert William, 82
Wright, Edmund Gordon, 82

Auxiliary Fire Service Firefighters, 80, 82, 83

*Avocet,* 77

Awards
American Merchant Marine Seamanship Trophy, 71
Auxiliary Fire Services, 82
Ray, Henry Albert William, 82
Wright, Edmund Gordon, 82
9 Crewmen on *Fire Fighter* presented Gallant Ship Unit
Citations and Ribbon Bars, 109
*Fireboat 5* firemen rescue of *Sansinena* crewmen awarded
Medals of Valor, *113,* 114
Honorary Fire Chief of DCFD to John H. Glenn, *100*
Jersey City firemen for heroism, 85
Lieut. James F. McKenna — Merchant Marine Meritorious
Service Medal, 109
Naval Distinguished Service Medal Award to Sub-officers,
82
New York fireboat fire fighters for heroism, 85
*Sea Witch* and *Esso Brussels* crewmen in rescue work, 109
U.S. Dept. of Commerce Gallant Ship Award to *Fire Fighter*
crew in 1975, 109

Ball, Walter, L., 114

Baltimore boats
*D'Alesandro, Mayor Thomas,* 94, 95
*Emrich, August,* 100
*Grady, Mayor J. Harold,* 100
*Torrent,* 63
*Wilkinson, P.W.,* 100

Baltimore fires
1904 — Great Baltimore Fire, 26, *26*
1936 — Norwegian freighter, *Gisla,* 63

Band, FDNY, 65

Battalion 6 Commander
Burmester, Bill, 4

Battalion Chief
Douglass, Jack S., 106
Gifford, Bethel F., 105
Gross, Harry, 89
Kenlon, John, 26
Strong, Arthur, *104*

*Battle, James,* 82

Battleship Row, 74, 75, 76, 77

Battleships
    *USS Arizona*, 74, *74*, 75, 76, 77, 78

Baumgartner, Henry J., 89
    Fire Chief, died in line of duty in 1947, 89

Beatteay, Bob, 105

Belgium boat
    *Esso Brussels, 108,* 109

Benevolent Association of the Paid Fire Dept. of Chicago, 34

Bertelsen & Petersen
    Builders of fireboat *Dowd, 37, 82*

Berths
    Brooklyn berth, 26
    East River berth, 24
    Berth 3, 76
    Berth 29, 114
    Berths 45-47, 109
    Berth 46, 109
    Berth 55, 114
    Berth 73, 2,3 *5, 7*
    Berth 85, 2
    Berths 103-110,*86*
    Berth 151, 85
    Berth 166, 90
    Berth 174, 106, *107*
    Berth 199, 100
    Berth 200A, 100
    Berth 223, 85

Bethlehem Shipbuilding Corp., 65

Bethlehem Steel, 45

Big Bertha — Cannon gun, *48*, 51, 52, *62,* 85, 100

Bikini, Atomic bomb tests, 77

Blackfriars Bridge, 82

*Blankenburg, Rudolph,* 94

Boats
    *Admiral Farragut,* 40
    *Aeolian* — Los Angeles, 45, *45*
    *Alki* — Seattle, 56, *56,* 58, 59, 115, *115,* 116
    *America* — New York ocean liner
    *Amphibian* — Milwaukee, 115
    *Anna Marie* — Los Angeles, 4, 8
    *Athlon,* 40
    *Avocet*
    *Battle, James* — Detroit, *82*
    *Blankenburg, Rudolph* — Philadelphia, 94
    Boston Whaler, 69
    *Bremen,* 24, 25, 26
    *Brennan, William J.* — Nework, *87*
    *Brooks, John M.* — Boston, 20, *21,* 38
    *Busse, Fred A.,* — Chicago, 62, *64,* 83, 97
    *Callahan, John H.,* — Florida, 43, 44

*Campbell, David* — Portland, Ore., *53*
*Cataract* — Milwaukee, 22, 26, *26, 43*
*Celebreeze, Anthony J.* — Cleveland, *91*
*Challenger, 118*
*Chicago* — Chicago, 33
*Chief Seattle* — Seattle, 115, *115,* 117, 118, 110
*City of Miami,* 114
*City of Oakland (Port of Oakland) (Hoga)* — Oakland, 78, 79, *80,* 100, *117*
*City of Portland* — Portland, Me., *100*
*Clevelander* — Cleveland, *23*
*Commencement* — Tacoma, 119, 123, 124
*Cotter, Edward M. (Grattan, W.S.)* — Buffalo, 60, *60*
*Crown Odyssey* — Cruise liner, 119
*D'Alesandro, Mayor Thomas* — Baltimore, 94, 95
*Defiance* — Tacoma, 119, *120,* 123, 124
*Delaware,* — Philadelphia, *93,* 94
*Deluge* — Milwaukee, *94,* 94
*Deluge* — New Orleans, 45
*Detroiter,* — Detroit, 21
*Dowd, John P. (Engine 47)* — Boston, *37, 55, 82*
*Duane, James* — New York, *28,* 29, 45
*Duffy, Michael P.* — Newark *87*
*Duwamish* — Seattle, 38, *38,* 39, *39,* 40, *40,* 41, *41,* 56, 115
*El Estero* — Panama, 84, *84,* 85, *85,* 99
*Emil deChamp* — France, 81
*Emrich, August* — Baltimore, *100*
*Engine 7* — Portland, Me., 37
*Esso Brussels* — Belgium — 108, *108,* 109
*Falcon,* — Los Angeles, *20*
*Farley, John H.* — Cleveland, *23*
*Fireboat No. 1* — Los Angeles, 4, 6, *6,* 45, *45,* 51, *61, 62, 104,* 121, 122, *122,* 124
*Fireboat No. 2 (Scott, Ralph J.)* — Los Angeles, 2, *3,* 4, *4,* 8, 45, *46,* 47, *48,* 49, *49, 50, 50,* 51, 85, *86,* 90, *90,* 91, *97,* 100, *101,* 103
*Fireboat No.3* — Los Angeles, 4, *84,* 85, 90, *97, 102,* 104
*Fireboat No. 4 (Gifford, Bethel F.)* — Los Angeles, 4, *5,* 6, *6, 104,* 105, 106, *107, 112,* 114
*Fireboat No. 5* — Los Angeles, 6, *6,* 8, 106, *113,* 114
*Fireboat No. 110* — Los Angles County, *111*
*Fire Fighter (Fighter)* — New York, 65, *65,* 66, *67,* 67, *68, 69,* 69, *70,* 71, *71,* 83, 84, 85, *85,* 98, *102,* 108, 109, *109,* 116
*Flanders, William L.* — Boston, 9, *10,* 20, *21*
*Foley, James,* 23
*Franklin, Benjamin (Franklin)* — Philadelphia, *93,* 94
*Fredericksburg,* — Los Angeles, 85, 114
*Fuller, John* — New York, 21
*Gaynor, Willilam J.* — New York, *44,* 65
*General Slocum* — Hoboken, N.J., *24,* 25, 26
*Gifford, Bethel F. (Fireboat No. 4)* — Los Angeles, 4, *5,* 6, *6, 104,* 105, *107, 112,* 114
*Gisla* — Norway, 63
*Glenn, John H. Jr.,* — New York, Washington D.C., *100*
*Governor, Alfred E. Smith* — New York, *125*
*Governor Irwin* — San Francisco — *30*
*Governor Markham, 30*

*Grady, Mayor J. Harold* — Baltimore, *100*
*Grandcamp*, 89, *89*
*Grattan, W.S. (Cotter, Edward M.)* — Buffalo, 35, 36, *36*, *56*, 58, 59, *59*, 60, *60*
*Guardian* — Boston, *114*
*Harvey, John J.* — New York, 62, 65, *72*, *84*, 85
*Havemeyer, William F.* — New York, 16, 19, 21
*Hawaiian Rancher* — Stockton, Ca., 79, *79*, 100
*Hewitt, Abram S.* — New York, 26, *27*, *71*
*High Flyer*, 89, *89*
*Hoegh Mallard* — Tacoma, 124
*Hoga (CIty of Oakland) (Port of Oakland)* — Pearl Harbor, 74, *74*, 75, 76, 77, 78, *78*, *117*
*Hull 856* — New York, 65
*Hutchinson, John M.* — Buffalo, 9
*Illinois (Roen, John III)* — Chicago, 33
*Kaiser Wilhelm der Gross*, 24
*Kettner, Bill*, *42*, 43
*Laila* — Boston, 63
*Lawrence, Cornelius W.,* — New York, 29
*Lee, Sally*, *100*
*Liberty*, *118*
*Liberty Ships*, 64
*Low, Seth* — Brooklyn, 12, 13
*Luna, Joseph*, *87*, *104*
*Mackenzie, William Lyon*, *102*
*Main*, 24, 25
*Markay* — Los Angeles, 90, *90*, 91, 114
*Massey Shaw* — London, 80, 81, *81*, 82, 83
*McClellan, George B.*, 12, *13*, *98*, 99
*McColl*, 58
*McDonald* — Boston, *22*, *55*, *63*
*McGiven* — New Orleans, 83
*McGonagle, William A.* — Buffalo, 59, *59*
*McKean, John D.* — New York, 94, 98, *98*, 99, *99*, *111*, 119
*McTighe, James (USS Bulwark)*, 88, *88*
*Medill, Joseph (I)* — Chicago, 34, *36*, 44, *64*, 94
*Medill, Joseph* — Chicago, 94, *96*, 97
*Mercy*, 80
*Mistress Delta*, 87
*Mitchell, John Purroy* — New York, *44*
*Monitor,* — Virginia, 17
*Moore, J. Hampton* — Philadelphia, *53*, 94
*New Yorker* — New York, 9, *10*, *12*, 12, 16, 17, 18, 24, *24*, 26, 27, *44*
*Newell, Harry,* — Ketchikan, *117*
*Normandie* — France — 83, 84
*Nortrans Visions* — Los Angeles, *111*
*Ocean Steelhead* — Tacoma, 124
*Oglala*, 76
*"Old Ironsides",* — Boston, *116*
*Orient Trader* — Greek, *102*
*Osborn, Lewis* — Boston, 20
*Phoenix* — San Francisco, *80*, 94, *95*, 97, *97*, 98, 103
*Polanic* — Yugoslavia — *94*
*Port Houston* — Houston, 45
*Porto Rican*, *117*

*Potter, George R.* — Buffalo, 9, *11*, 58, 59, 64
*Ring, Thomas* — Boston, 55
*Roen, John III (Illinois)* — Chicago, 34
*Rolls-Royce* of fireboats — New York, 62
*Saale*, 24
*Samuel, Bernard* — Philadelphia, *92*, 93, *93*, 94
*Sansinena* — Liberian, 109, *112*, 113, *113*, 114
*Scannell, David* — San Francisco, 30, *31*, *32*, 33, 97
*Schlaeger, Victor L.* — Chicago, *94*, 97
*Scott, Ralph J. (Fireboat No. 2)* — Los Angeles, 2, 4, 6, 8, 45, *47*, *48*, 51, 51, 52, *52*, *54*, 56, *56*, 59, 67, *86*, 106, *107*, *111*, *113*, 114
*Sea Witch*, 108, *108*, 109, *109*
*Silver Ash* — Brooklyn 71
*Smoke I* — New York, *87*
*Smoke II* — New York, *98*
*Snoqualmie* — Seattle, 20, *21*, 40, 41
*SS Muenchen* — New York, *72*
*Stewart, Graeme* — Chicago, 34, *36*, 44
*St. Florian* — Boston, *112*
*Strong, William L.* — New York, 24, 26
*Stuart, Edwin S.* — Philadelphia, *23*, 94
*Sullivan, Dennis T.* — San Francisco, 30, *31*, *32*, 33, *86*, 97
*"Super Fireboat"* — New York, 12
*Swenie, Denis J. (Geyser)* — Chicago, 33, 34, *36*
*The Floating Engine* (known as *The Floater or Engine Company No. 42)*, 17, 18
*Torrent* — Baltimore, *63*
*United States* — New York, 64
*USS Arizona*, 74, *74*, 75, 76, 77, 78
*USS Bulwark (McTighe, James)* — Boston, 88, *88*
*USS California*, 75
*USS Constitution* — Boston, *116*
*USS Nevada*, 74, *74*, 75, 76, 77, *77*, 78, 79
*USS Oklahoma*, 75
*USS Shaw*, 76
*USS Utah*, 75
*USS Vestal*, 75
*USS West Virginia*, 75, 77
*Virginia* — Virginia, 17
*Warrior* — Los Angeles, *20*
*Wilkinson, P.W.* — Baltimore, *100*
*Willett, Thomas* — New York, *28*, 29, 72, *91*
*Williams, George H.* — Portland, Oregon, 29, *29*
*Yosemite (Protector) (Conway, Michael J.)* — Chicago, 22
*Zophar Mills* — New York, 16, 24, 26, *27*

Boston boats
*Brooks, John M*, 20, *21*, 38
*Dowd, John P. (Engine 47)*, *37*, 55, 82
*Flanders, William L.* 9, *10*, 20, *21*
*Guardian*, 114
*Luna, Joseph*, *87*, *104*
*McDonald*, 22, 55, 63
*McTighe, James F.*, 87
*Ring, Thomas*, 55
*St. Florian*, 112

USS Bulwark (McTighe, James), 87, 88, *88*
USS Constitution, *116*

Boston fireboat fleet, *88*

Boston fires
   1872 — Mercantile district, 20
   1911 — Warren Coal Co., 37
   1933 — Boston's Central wharf, *22*
   1937 — Freighter *Laila*, 63
   1961 — Castle Island Terminal, 104
   1984 — Fire on pier, *116*

Boston Whaler, 69

Boston Whaler motorboat, 69, 119

Boston Whalers Inc.,
   Supplier of small fireboats, *114*

Bowes, Thomas D., 92, 96, *100*

Braidwood, James, *16*, 17
   Died in line of duty in 1861, *16*, 17

Braithwaite, John 17

*Bremen*, 24, *25*, 26

Bresnan nozzles, *7*, 8, 103

Bridges
   Peace, 59
   Verrazano-Narrows, 71, 108

British
   Admiralty, 80
   Metropolitan Air Force, 79
   Navy guns, 81
   Troops, 80

Bronx, New York, 26

Brooklyn boats
   *Low, Seth*, 12, 13
   *Silver Ash*, 71

Brooklyn fires
   1939 — Abroad *Silver Ash*, 71
   1956 — Luckenback steamshp pier, 71
   1960 — Aircraft Carrier *Constellation*, 71
   1962 — Sinclair Refining Co., *125*
   1969 — 27th Street pier, *99*

*Brooks, John M.*, *20*, *21*, 38

Brown, Bob, 74, 75, 77, 78

Brussels ports, 109

Buda Company, 99

Buda diesels, 78

Buddy system (skin diving), 6

Buffalo boats
   *Cotter, EdwardM. (Grattan, W.S.)* 36, 56, 58, 59

   *Grattan, W.S. (Cotter, Edward M.)*, 35, 36, *36*, *56*, 58, 59, *59*, 60, *60*
   *Hutchinson, John M.* 9
   *Potter, George R.*, 9, *11*, 58, 59, 65

Buffalo City Hall, 58

Buffalo Drydock Company, 59

Burlington and Armour & Co., 33

Burlington Railway Company, 33

Burmester, Bill, 4

*Busse, Fred A.*, 62, *64*, *83*, *96*

Byron Jackson centrifugal pumps, *46*, 49, *49*, 51, 56, 57

California Fire Service Day, 105

*Callahan, John H.*, 43, 44

Calverly, L.E., *46*, 49

Cameo Program and Computer System, 119

Camouflage gray, 83
   Fireboats along
      Atlantic Ocean
      Pacific Ocean
      Gulf coasts

*Campbell, David*, *53*

Captains
   Allen, Maurice, 85
   Columbi, Richard A., 116
      Deputy Chief of Seattle Fire Department, 116
   Gordon, Jack, 91
   Gorter, Henry J., 118
      Master mechanic, 118
   Hibbard, Don, 2, 4, 6, 8
      Maritime Marine Division Commander, 2, 4, 6, 8
   Lawrence, Warner L., 51, 52, 54
   Van Schaick, William H., 26

Carson, Pirie, Scott & Co., 34

Carter, Jimmy, 79

Casco Bay Islands, *100*

Castle Park, *24*

*Cataract*, 22, 26, *26*, *43*

Caterpillar D333, 78
   3208 Caterpillar diesel engines, *111*
   3160 Caterpillar Waterous pump, *111*

Cave Point, 84

*Celebreeze, Anthony J.*, *91*

*Challenger*, *118*

*Chicago*, 33

Chicago City Hall, 20

Chicago boats
  *Busse, Fred A.*, 62, *64, 83, 96*
  *Chicago,* 33
  *Illinois (Roen, Jon III),* 33
  *Medill, Joseph,* 34, *36,* 44, *64,* 94
  *Medill Joseph,* (new boat, same name), 94, *96,* 97
  *Roen, John III (Illinois), 34*
  *Schlaeger, Victor L., 94,* 97
  *Stewart, Graeme,* 34, *36,* 44
  *Swenie, Denis J. (Geyser),* 33, 34, *36*
  *Yosemite (Protector) (Conway, Michael J.),* 22

Chicago Fire Insurance Patrol, *96*

Chicago fires
  1871 — Great Chicago Fire, 19, 20, 30, 35
  1908 — Great Grain Elevator fire, 33, 34, 35, *36*
  1939 — Elevators — Norris Grain Co., *64*
  Postwar, Lumber yard, 83
  1950 — Container Corporation Warehouse, *95*
  1951 — Warehouse and Light Manufacturing Plant, *96*
  1957 — Rock Island Grain Elevator, *95*
  1966 — Box factory, *62*
  1967 — McCormick Place, 97

Chief Boatswain's Mate
  McManus, Joe B., 75, 76, 77, 78

*Chief Seattle,* 115, *115,* 117, 118, 119

Choner, Jim, 2, 4

Christened boat
  Lee, Sally christened *Lee, Sally,* 100
  Flanagan, Eleanor Grace, christened *Hull 856,* 65
  Scott, Mrs. Ralph (Adeline)
    Christened *Los Angeles City No. 2,* 49, 52

Christy Corporation, 94

Chrysler Corporation Marine engines, 106

Churchill, Winston, 80, 81

*City of Miami, 114*

*City of Oakland (Port of Oakland) (Hoga),* 78, 79, *80,* 100, *117*

*City of Portland,* Portland, Me., *100*

Cities
  Atlantic Beach, Fla. *100*
  Baltimore, 9, 19, 26, 30, *63, 96*
  Bay City, Mi., *62*
  Boston, Mass., 9, 19, 44, *63, 87,* 88, *95, 100,* 104, *112, 114, 116*
  Bremerton, Wa., 77
  Brooklyn, 9, *99,* 108, *125*
  Buffalo, 9, 21, 58, 59, *60*
  Camden, N.J., *28,* 92, *94, 96,* 98
  Chicago, Ill., 9, 30, 34, 35, 39, 44, *62, 64,* 94, *95, 96,* 97
  Cleveland, 9
  Coddington, R.I., *69*

  Detroit, Mi., 9, 21, *82*
  Elizabeth, N.J., 35
  Elmira, N.Y., 9
  Escatawpa, Miss., *118*
  Galveston, Texas, 89, *90*
  Hampton Roads, Va., 17
  Hoboken, N.J., 24, *25,* 30
  Hudson, N.Y., 9, 13
  Jersey City, N.J., 85
  Los. Angeles, *20,* 45, *47,* 49, *50,* 51, 52, *54,* 56, *84,* 85, *97, 101, 104, 105, 106, 111, 118, 120*
  Manchester, N.H., 9
  Milwaukee, 9, *23,* 94, 115
  Morris Heights, N.J., 75
  New Haven, Conn., 20, *100*
  New Orleans, *45*
  New York, 9, 19, 26, 30, 35, 39, 45, 62, 64, 65, *66, 68,* 69, 71, 84, 85, *98, 100, 102,* 105, 108, 109, *125*
  Norfolk, Va., *100*
  Oakland, Ca., 56, *79,* 100
  Oxnard, Ca., *106*
  Philadelphia, Pa., 9, 19, *53,* 92, *93,* 94, *111*
  Portland, Oregon, 29, *53*
  Rochester, New York, 21
  San Diego, 41, *42,* 43, *98*
  San Francisco, 29, 30, *30, 31,* 35, 39, 94, *95,* 97, 103, 105, *110,* 118
  San Pedro, *20,* 45, 52, *97*
  Seattle, 9, 20, 38, *56,* 115, 116, 117, 119
  Seneca Falls, N.Y., 9
  Sturgeon Bay, Wis., 94
  Tacoma, Wa., *62,* 105, 115, *120,* 122, 123, 124
  Texas City, Texas, *89,* 91
  Washington, D.C., *100*
  Wilmington, Del., *20,* 45, 90
  Wisconsin Rapids, Wis., 34

Civil War, 9, 17, 18, 19, 21

Clapp & Jones, 9, *13,* 33
  Manufactured pumps, 9, *13,* 33
  Steam engine builder, 9, *13,* 33

Cleveland boats
  *Celebreeze, Anthony J.,* 91
  *Clevelander,* 23
  *Farley, John H.* 23

Cleveland fires
  1984 — Waterfront blaze, *23*

*Clevelander,* 23

Coast Guard, 84, *87,* 92

Coast Line Shipbuilding Company, *62*

Cockerell, Sir Christopher, 123

Colman Dock and its warehouse, 40

Columbi, Richard A., 116

Columbian Bronze Company, 97

Commander
  Ramaekers, Gerald, 2

*Commencement,* 119, 123, 124

Commencement Bay, 120, *120*, 123

Consolidated Papers, Inc., 34

Consolidated Shipbuilding Corporation, 75

Construction
  Ericsson, John, 17
  Vosper, Hovermarine, 123

Continental Grain Company, *95*

Coolidge Propellers, 117

Cooper, Patrick, 40, 41
  Died in line of duty in 1914 Grand Trunk Complex fire, 41
  While on duty on *Duwamish*

Corcoran, Mike, 4

Corrigan, Michael J., *96*

*Cotter, Edward M., 60*
            President of Buffalo Fire Fighters Local 282

*Cotter, Edward M. (Grattan, W.S.),* 60, *60*

Cowles, William C., 9, 12, 16, 17, 18, 92

Crewman
  Ray, Henry Albert William, 82
  Wright, Edmund Gordon, 82

Crescent Shipyard, 35

Croker, Edward F., 24, 26, *27,* 28, *28,* 29, 34, 39, 105

*Crown Odyssey,* Cruise liner, 119

Cummins Engine Company, 97, 98, 105

Dahlquist, Bill, 2, *2,* 8

Damrell, John S., 20

*D'Alesandro, Mayor Thomas,* 94, 95

Data banks, 119

Darley bronze pump, *114*

Dean-Hill centrifugal pumps, *62,* 97

*Defiance,* 119, *120,* 123, 124

Defoe Shipbuilding Company, *62, 94*

De Laval pumps, *62, 66,* 67, *100,* 105, *116*

De Laval Steam Turbine Company, 41

De Laval Turbine Pacific Company pumpts, 98

*Delaware, 93, 94*

Delco Products, 99

*Deluge,* 45

*Deluge, 94, 94*

Denver's Regis College, 89
  Earthquake center, 89

Deputy Chief
  Columbi, Richard A., 116
  Mork, Stan, 122
    Deputy Chief of Harbormaster, 122

Destroyer boat
  *USS Shaw,* 76

Detroit Allison diesel engines, 117

Detroit boats
  *Battle, James, 82*
  *Detroiter,* 21

*Detroiter,* 21

Died in line of duty
  1861 — James Braidwood, *16,* 17
  1906 — Dennis T. Sullivan 30, *31, 32,* 33, 97
  1911 — David Campbell, *53*
  1914 — Patrick Cooper, 40, 41
  1928 — *Grattan* engineer, Thomas Lynch, 58
  1930 — John J. Harvey, 62, 65, 72, *84,* 85
  1941 — Bombardier Noburo Kanai, 75, 78
  1947 — Henry J. Baumgartner, 89
  1953 — John D. McKean, 12, *13, 98,* 99
  1963 — Arthur Strong, *104*
  1973 — John L. Paterson, 108
  1986 — Harry Newell, *117*

*Diesel Shipbuilding Company, 100*

Ditzel, Paul, 105

Dock 1010, 76

Douglas, Owen, 122

Douglass, Jack S., 106

*Dowd, John P. (Engine 47), 37, 55, 82*

Drake Craft, Inc. *106*

*Duane, James, 28,* 29, 45

Duluth, Missabe & Northern Railway, 59

Dunkirk evacuation
  1940 — Explosions of shells, 80, 81

Dunkirk (Operation Dynamo), 79, 80, 81

*Duwamish,* 38, *38,* 39, *39,* 40, *40,* 41, *41,* 56, 115

Eagle Engine Company No. 13, 16

*El Estero,* 84, *84,* 85, *85,* 99

Elliott Bay Design Group, 56, *56*

Ellis Island, 24, *24*

mblems, 75
  Rising Sun on Japanese bombers, 75

*nil deChamp*, 81

*nrich, August, 100*

ncinal Terminal, Alameda, 79

*ngine 7,* 37

ngine 8, Buffalo firemen, *60*

ngine 20, Buffalo *60*

*ngine 47 (Dowd, John P.),* 37

ngineers
  Choner, Jim, 2, 4
  Lynch, Thomas, 58
  Mantor, George, 56
    Chief engineer, 56
  McKean, John D., 12, *13, 98,* 99, *99*
    Marine engineer, 12, *13, 98,* 99
  Rasmussen, John, 2, 14
  Reddan, Danny, *66*
    Marine engineer, *66*

ngland fire apparatus maker, *14,* 15
  Shand Mason & Co., *14,* 15

nterprise Engine and Machinery Company, 99

quitable Equipment Company, *98*

ricsson, John, 17

*sso Brussels, 108,* 109

xport-Import Centers, 19
  Baltimore
  Boston
  New York
  Philadelphia

airmont Hotel, 30, *30*

*alcon,* 20

*arley, John H.,* 23

DNY Disaster Unit, 109

ellows and Stewart Division of Harbor Boat Building Co., 52

erquson Propeller and Reconditioning Company, 98

ireboats, See Boats (Insert)

ire Box No. 15, 90

ire Chief
  Alderson, John H., 85, 91
  Algren, Louis, 43
  Baumgartner, Henry J., 89
  Croker, Edward F., 24, 26, *27,* 28, *28,* 29, 34, 39, 105
  Dasmrell, John S., 20
  Hill, Raymond M., 51, 52

Johnston, William, Jr., 103, 104, 105
  Assistant Chief, 103, 105
Mitchell, Tony F., 105, 122, 123, 124
Reiser, James W., 122
Scott, Ralph J., *46,* 49, 51
Shaw, Sir Massey Eyre, 80, *81*
Smith, John E., *100*
Stetson, Frank L., 40
Williams, Robert A., 19

Fire Commissioners
  Corrigan, Michael J., *96*
  Hedden, George W., 58, 59, *59*
  McElligott, John J., 65
  Quayle, Frank, *91*

Fire Departments
  Baltimore Fire Dept., *96*
  Boston Fire Dept., 45, *88,* 104
  Buffalo Fire Dept., 35, 39
  Chicago Fire Dept., 34, 35, 39
  District of Columbia Fire Dept. *100*
  Alaska Ketchiken, Fire Dept., *117*
  Los Angeles County Fire Dept., *111*
  Los Angeles Fire Dept. 2, *5,* 6, 6, 8, 45, 49, 51, 85, 89, *102,*
    103, 104, 105, 106, *106,* 114
  New York Fire Dept., 16, *18,* 22, 26, 29, *44, 70,* 71, *72, 85,*
    *100, 102,* 108
  New York Volunteer Fire Dept., 16, 17
  Oakland Fire Dept., 79
  San Diego Fire Dept., 43
  San Francisco Fire Dept., 118
  Seattle Fire Dept., 40, 56, 116
  Tacoma Fire Dept., *61, 104,* 119, 122, 123, 124
  Texas City Volunteer Fire Dept., 89

"Fire Engineering" 105

Fire horses in Chicago, 34, 43

Fire Marshall
  Horan, James, 30, 33, 34

Fire Station move, 79
  *City of Oakland* to new berth in Jack London Square, 79

*Fireboat No. 1,* 4, 6, *6,* 45, *45,* 51, *61, 62, 104, 121,* 122, *122,* 124

*Fireboat No. 2 (Scott, Ralph J.,), 2, 3,* 4, *4,* 8, 45, *46,* 47, *49, 50,*
  51, 85, 90, *90,* 91, *97,* 100, *101,* 103

*Fireboat No. 3, 4, *84,* 85, 90, *97, 102,* 105

*Fireboat No. 4 (Gifford Bethel F.),* 4, 5, 6, *6, 104,* 105, 106, *107,*
  *112,* 114

*Fireboat No. 5,* 6, *6,* 8, 106, *113,* 114

*Fireboat No. 110. 111*

Fireboat fleet of Chicago 35

*Fire Fighter (Fighter),* 65, *65,* 66, 67, *67, 68,* 69, *69, 70, 71,* 71, 83,
  84, 85, *85,* 98, *102,* 108, 109, *109,* 116

Firemen
   Beatteay, Bob, 105
     Skin diver 105
   Cooper, Patrick, 40, 41
   Hylant, John, 58
   Kemperman, John, 114
   McDaniel, John, 7, 8
   Roquemore, Joseph V. "Rocky", *84*, 85
   Smith, "Buzz", 4
   Stokes, John W., 40
   Taylor, Forrest E., 114
   Thorson, Glenn, 4
   Vidovich, Frank, 4

F.N. McIntyre Brass Works, 94, 98

Fisherman's Cooperative Association Building, 4

Fishing boats, 8

Fitzsimmons, Matthew T., Jr., 109

Flanagan, Eleanor Grace, 65
   Christened *Hull 856,* 65

Flanagan, Joseph, 65

*Flanders, William L,* 9, *10,* 20, 21

Florida boat
   *Callahan, John H.* 43, 44

*Foley, James,* 23

Founder
   Reid, Daniel G., *111*
     Founder of U.S. Steel Corporation, *111*

*Franklin, Benjamin, (Franklin),* 93, 94

*Fredericksburg,* 85, 114

Freighter boats
   *Admiral Farragut,* 40
   *Laila,* 63
   *Ocean Steelhead,* 124
   *Orient Trader,* 102
   *Polanic,* 94
   *SS Muenchen,* 94

French boats
   *Emil deChamp,* 81
   *Grandcamp,* 89, *89*
   *Normandie,* 83, 84

French freighters
   *Emil deChamp,* 81
   *Grandcamp,* 89, *89*

French ocean liner
   *Normandie,* 83, 84

Fridell, William, *62*

*Fuller, John,* 21

Gailbraith — Bacon Dock and Warehouse, 40

Gallant Ship Unit Citations and Ribbon Bars, 109, *109*

Garfield, Harry A., 41

*Gaynor, William J., 44,* 65

G E M, 123
   Ground-effect-machine, 123

General Electric Curtis steam turbine, 29, 30, 31

General Electric diesel engines, 94

General Electric generators, 105

General Motors, 56, 58
   Diesels, 56, 58

General Motors, *66,* 67
   Winton/Cleveland, *66,* 67

General Motors Detroit Marine Engine, 124

General Motors engine, 94
   Rockford Clutch, 94

General Motors series, 93, *116*
   Diesel engines

General Ship & Engine Works, *88*

*General Slocum, 24,* 25, 26

German Artillery and Aircraft, 80

German Land Installations, 81

German plane, 81
   Dropped bomb

Gibbs, William Francis, 64, 65, 67, 69, 71, *102*

Gifford, Bethel F., 105

*Gifford, Bethel F. (Fireboat No. 4),* 4, 5, 6, *6, 104,* 105, *107, 112,*
   114

*Gisla,* 63

Gislow, Henry, Inc., *72*

*Glenn, John H., Jr., 100*

Glennfer, diesel engine, 80

Gold Room, 67, 109

Gold Rush, 16, 29, 39

Golden Gate, *117*

Gordon, Jack, 91

Gorter, Henry J., 118

*Governor Alfred E. Smith, 125*

*Governor Irwin, 30*

*Governor Markham, 30*

Gow., Paul A., 116, 118

Grace Lines Terminal Building, *91*

*Grady, Mayor J. Harold,* 100

Grand Trunk Complex, 40

*Grandcamp,* 89, *89*

*Grattan, W.S. (Cotter, Edward M.),* 35, 36, *36, 56,* 58, 59, *59,* 60, *60*

*Grattan* engineer, *56*
  Died in line of duty in 1928
    Oil barge explosion and fire, *56*

Gray, Brainard "Choppy", 85, 90, 91

Graveyard Bay, 109

Graveyard of Ships, 80

Great Depression, 58, 59, 62

Great Lakes cities, 59

Great Lakes fireboats, 35, *62,* 94

Great Lakes freighters, 33
  Fire, 33

Great Lakes ports, *91,* 92
  Access to Atlantic Ocean, 91
  Access to sea and foreign commerce, 92

Great Lakes vessels, 59

Gross, Harry, 89

*Guardian,* 114

Gulf of Mexico, *98*

Hageman, Edward C., 116

Halifax, Nova Scotia, *82*

Hand operated fireboats, 8

Hanson, Hans, 34

Harbor Grain Terminal, 106, *107*
  Berth 174, 1967 fire, 106, 107

Harbormaster, deputy chief, 122
  Mork, Stan, 122

Harvey, John J., *72*
  Pilot of *Thomas Willett*
  Died in line of duty in 1930, *72*

Harvey, John J., 62, 65, *72, 84, 85*

*Havemeyer, William F.,* 16, 19, 21

Hawaii, 78

*Hawaiian Rancher,* 79, *79,* 100

Hedden, George W., 589, 59, *59*

Hell Gate, 26

*Hewitt, Abram S.,* 26, *27, 71*

Hibbard, Don, 2, 4, 6, 8

Hickman and Wheeler Fields, 75

*High Flyer,* 89, *89*

Hill, Raymond M., 51, 52

Hilzer, Charles C., 106

H. Newton Whitteley, Inc., *100*
  Designer of *Glenn, John H., Jr., 100*

Hoboken New Jersey boat
  *General Slocum, 24,* 25, 26

Hoboken, N.J. fire
  1900 — North German Lloyd Co. Terminal, 24, *24, 25,* 26

*Hoegh Mallard,* 123

*Hoga (City of Oakland) (Port of Oakland),* 74, *74,* 75, 76, *77, 77,* 78, *78, 117*

Honorary Fire Chief of DCFD, *100*
  Glenn, John H., *100*
  See Awards

Horan, James, 30, 33, 34

Houston boat
  *Port Houston,* 45

Hovercraft, 123
  See Vosper, Hovermarine, 123

Hovermarine, Vosper, 123

*Hull 856,* 65

Humber Bay, *102*
  Near Ward's Island, *102*

Humphrey's Warehouse, 17
  On River Thames

*Hutchinson, John M.,* 9

Hylant, John, 58

Hylant, Thomas, 58

*Illinois (Roen, John III),* 33

Imperial Company generators, 98

Invasions, 77
  of France
  of Iwo Jima
  of Okinawa

Inventors
  Braithwaite, John, 17
  Cockerell, Sir Christopher, 123
  Maxim, Sir Hiram Stevens, 18, 19

Island, Oahu, 74

Jackson Shipyard, *100*

Jackson, Frank Whitford, 82

Japanese Task Force, 74

Jersey City Firefighters, 84
   Fire, 84, 85

Jersey City firemen
   See Awards for heroism, 85

John Deere generators, 117

John H. Mathis Company, 98

John H. Wells, Inc. of New York, 98

Johnston Outboard Motors, 114

Johnston, Roy G., 76

Johnston, William W. Jr., 103, 104, 105

*Kaiser Wilhelm der Gross*, 24

Kanai, Noburo, 75, 78

Kaneohe Naval Air Station, 75

Kemperman, John, 113

Kenlon, John, 26

Keough, William F., *37, 82*

*Kettner, Bill, 42*, 43

King George VI, 82

Korean War, 52

Korean War military traffic, 91

Kuhn, John, 122

LaFrance Manufacturing Co., 9

Laguardia, Firoello H., 64

*Laila*, 63

Lakes
   Lake Calumet in South Chicago, 97
   Lake Erie, 21, 35, 58, *60*
   Lake Superior, 34

Lake Washington Ship Canal, 41
   Chittenden Locks, 41

Lambeth Headquarters, 82
   London Fire Brigarde, 82

Landing C., Hospital Point, 76

Landmarks Preservation Committee, 62
   Ratcliffe, Dr. Allen W., 62

*Lawrence, Cornelius W.*, 29

Lawrence, Warner L., 51, 52, 54

L.C. Norgaard & Associates, 105

LeCourtney Pumps, *72*

Lee, Richard C., *100*

Lee, Sally, *100*

*Lee, Sally, 100*

Leece-Neville Alternators, 124

Lewis Machine Gun, 81

Liberian boat
   *Sansinena*, 109, *112*, 113, *113*, 114

*Liberty*, 118

*Liberty Ships*, 64

Lieutenants
   Flanagan, Joseph, 65
   McKenna, James F., 109
     Lieut. of Firefighters
   Schickenbantz, Henry, 58

Link Belt Company, 97

Locust Island, 26
   near Rikers Island in New York, 26

London boats
   *Massey Shaw*, 80, *81*, 81, 82, 83

London's East India docks, 83

London fires
   1861 — Great Tooley Street, *16*, 17
   1940 — Incendiary bomb fires, 82

London Fire Brigade officers
   Jackson, Frank Whitford, Commander, 82
   May, Aubrey John, 80, 82
     Sub-officer

London Mercantile Fire Brigade, 80, *81*, 82, 83

Long Beach, Ca., boats
   *Challenger*, 118
   *Liberty*, 118

Long Beach, Ca., offshore oil drilling platforms, *118*

Lords Commissioners of the Navy, 82

Los Angeles boats
   *Aeolian*, 45, *45*
   *Anna Marie*, 4, 8
   *Falcon*, 20
   *Fireboat No. 1*, 3, 4, 6, *6*, 45, *45*, 51, *61*, 62, *104*, 121, 122, *122*, 124
   *Fireboat No. 2 (Scott, Ralph J.)*, *2, 3*, 4, *4*, 8, 45, *46*, 47, 49, *49, 50, 51*, 85, 90, *90*, 91, *97*, 100, *101*, 103
   *Fireboat No. 3*, 4, *84*, 85, 90, *97, 102*, 105
   *Fireboat No. 4, (Gifford, Bethel F.)*, 4, 5, 6, *6*, *104*, 105, 106, *107*,. *112*, 114

*Fireboat No. 5,* 6, *6,* 8, 106, *113,* 114
*Fredericksburg,* 85, 114
*Gifford, Bethel F., (Fireboat No. 4),* 4, *5,* 6, 6, *104,* 105, 106, *107, 112,* 114
*Markay,* 90, *90,* 91, 114
*Nortrans Visions, 111*
*Scott, Ralph J. (Fireboat No. 2),* 2, 4, 6, 8, 45, *47,* 51, *52,* 52, *54,* 56, *56,* 59, 67, 106, *107, 111, 113,* 114
*Warrior,* 20

Los Angeles fires
    1944 — Navy landing craft and wooden dock, *84*
    1944 — Union Oil Terminal, 85
    1947 — Waterfront Calamity, 89
    1947 — Tankship *Markay* exploded, 90, *90,* 91
    1960 — Under Matson Lines wooden wharf, 100
    1960 — Matson Terminal, *101*
    1967 — Harbor Grain Terminal, 106, *106, 107*
    1974 — *Nortrans Visions* fire, *111*
    1976 — Oil tanker explosion, *112,* 113, *113,* 114
    1988 — Berths 73 and 85, 2, 3, 4, *5,* 6, 7, 8

Los Angeles Fire Department
    Paramedic Medical Command Post, 114

Los Angeles harbor, 103

Los Angeles International Airport, *111*

Los Angeles Shipbuilding and Drydock Corp. *47,* 49, *86*
    Todd Shipyard in San Pedro, *47*

*Low, Seth,* 12, 13

Lumber Port, 20

Luna, Joseph, *87, 104*

*Lynch, Thomas, 58*

*1925 Mack Hose wagon, 50,* 51

*Mackenzie, William Lyon, 50,* 51

Maddock, John, 122

*Main,* 24, *25*

Manhattan — Brooklyn Tunnel, 24

Manhattan Island, *111*

Mantor, George, 56

Margate Coast, 81
    Vicinity of Dunkirk, 81

Marina del Rey craft harbor, *111*

Marine Air Base at Ewa, 75

Marine Division of FDNY, 26, *44,* 62, 65, *68,* 69, 71, *72, 73, 85, 87, 91, 98, 99, 111*
    John Kenlon commander and Battalion Chief, 26

Marine wiper, 66
    McCrossen, Arty, 66

Maritime expertise, 122
    Douglas, Owen, 122

*Markay,* 90, *90,* 91, 114

Maslin, Marsh, 97

Mass Port Marine, *118*

*Massey Shaw,* 80, 81, *81,* 82, 83

Massey Shaw and Marine Vessels Preservation Society, Ltd., 83

M. Greenburg & Sons, 99

Master of Ship *Sea Witch,* 108
    Paterson, John L., 108

Mates
    Ball, Walter L., 114
    Corcoran, Mike, 4
    McManus, Joe B., 75, 76, 77, 78
        Chief Boatswain's mate, 75, 76, 77, 78
    Minehart, Marion, 75
        Chief machinist's mate, 75
    Planagan, John, 90, 91
        First mate, 90, 91
    Stoddard, Fred, 3, 6
        Supervising mate, 3, 6
    Svorinich, Mike, *6*

Matson Terminal, *101*

Maxim, Sir Hiram Stevens, 18, *19*

May, Aubrey John, 80, 82

MAYDAY call, 108

Mayors
    Jackson, Frank Whitford, 82
        Commander of London Fire Brigarde, 82
        Auxiliary Fire Services, 82
    Laguardia, Fiorello H., 64
    Lee, Richard C., *100*
    Schwab, Frank X., 58, 59
        Mayor of Buffalo, 58, 59

McAllaster, Eugene L., 39

*McClellan, George B.,* 12, *13, 98,* 99

*McColl,* 58

McCrossen, Arty, *66*

McDaniel, John, *7,* 8

*McDonald, 22, 55, 63*

McElligott, John J., 65

*McGiven,* 83

*McGonagle, William A.,* 59, *59*

McIntosh & Seymour diesel engines, 78

(American Locomotive), 78

McKean, John D., 12, *13*, *98*, 99
  1953 — Fatally scalded by boiler explosion, 12, *13*

*McKean, John D.*, 94, 98, *98*, 99, *99*, *111*, *119*

McKenna, James F., 109

McManus, Joe B., 75, 76, 77, 78

*McTighe, James (USS Bulwark), 87,* 88, *88*

Meade, Richard W., 17

*Medals of Valor, 91, 113,* 114
  See Awards

*Medill, Joseph,* 34, *36,* 44, *64,* 94

*Medill, Joseph,* 94, *96,* 97
  (New boat replaced by a new boat by same name)

Merchant Marine Meritorious Service Medal, 109
  See Awards

Merchant Ship Corp., 53

*Mercy,* 80

Michigan Avenue on Lake Front-Chicago, 34

Michigan Avenue Bridge, 59

Michigan Tool Company, 94

Milwaukee boats
  *Amphibian,* 115
  *Cataract,* 22, 26, *26, 43*
  *Deluge,* 94, *95*
  *Foley, James,* 23

Minehart, Marion, 75

Minelayer
  *Oglala,* 76

Mississippi River steamboat, 18

*Mistress Delta,* 87

*Mitchell, John Purroy, 44*

Mitchell, Tony F., 105, 122, 123, 124

*Monitor,* 17

Monsanto Chemical Company, 89
  Edgar M. Queen, board chairman, 89

*Moore, J. Hampton,* 94

Moran, McAllister Towing Companies, 109

Moran Towing Corporation, *119*

Mork, Stan, 122

Mosher water tube boilers, 39

Mount Rainier, snow-capped, 120

Mrs. O'Leary's Cow, 30
  Chicago fire

"20 Mule Team" United States Chemical Company, 90

National Association of Underwater Instructors, 103

National Board of Fire Underwriters, *62,* 83, 92

National Fire Protection Association, 83, *84,* 92

National Foam, 118

National Foam System, Inc., 98

National magazine writer, 105
  Ditzel, Paul, 105

National Oceanic and Atmospheric Administration, 119

National Park Service, *111*

National Register of Historic Places, *111*

Naval Architects
  Alden, John G., 94, *95,* 97, 98, 100
  Baalin, Fred A., 29
  Bowes, Thomas D., 92, 96, *100*
  Calverly, L. E., *46,* 49
  Cowles, William C., 9, 12, 16, 17, 18, 92
  Gibbs, William Francis, 64, 65, 67, 69, 71, *102*
    Marine architect, 64, 65, 67, 69, 71, *102*
  Gislow, Henry, Inc., *72*
  Hageman, Edward C., 116
    Chief naval architect, 116
  Keough, William F., *37, 82*
  McAllaster, Eugene L., 39
  Parsons, Harry deBerkeley, 9, 12, *13,* 16, 17, 21, *25, 27*
  Petrich, James, 124
  Sarin, Jack, *117*
  Stevens, Arthur D., 43, 44, 45
    Marine architect, 43, 44, 45

Naval Distinguished Service Medal, 82
  See Awards

Naval District Headquarters, 75

Navy and Marine Corps., 78

Navy Hospital Ship, 80
  *Mercy,* 80

Navy Photographer, 77

Navy Submariners, 65

Nelson, Jim, *120*

Nevada Point, 77

Newark boats, *87*

*New Haven boat*
  *Lee, Sally, 100*

New Orleans boats
  *Deluge,* 45

*McGiven,* 83

New York boats
 *America,* 64
 *Duane, James, 28,* 29, 45
 *Firefighter (Fighter),* 65, *65,* 66, 67, *67, 68,* 69, *69, 70,* 71, *71,*
  83, 84, 85, *85,* 98, *102,* 108, 109, *109,* 116
 *Fuller, John,* 21
 *Gaynor, William J., 44,* 65
 *Governor Alfred E. Smith,* 125
 *Harvey, John J.,* 62, 65, 72, *84, 85*
 *Havemeyer, William F.,* 16, 19, 21
 *Hewitt, Abram S.,* 26, *27,* 71
 *Hull 856,* 65
 *Lawrence, Cornelius W.,* 29
 *McKean, John D.,* 94, 98, *98,* 99, *99*
 *McKean, William D., 111, 119*
 *Mitchell, John Purroy, 44*
 *New Yorker,* 9, *10,* 12, *12,* 16, 17, 18, 24, *24,* 26, *27, 44*
 *Rolls-Royce* of fireboats, 62
 *Smoke II,* 98
 *SS Muenchen,* 72
 *Strong, William L.,* 24, 26
 *"Super Fireboat",* 12
 *United States,* 64
 *Willett, Thomas, 28,* 29, *72, 91*
 *Zophar Mills,* 16, 24, 26, 27

New York Bridge, 65

New York fires
 1879 — Greenwich Village, 21
 1904 — Hoboken disaster, 26, 27
 1908 — Central Railroad's Grain Elevator, 9
 1930 — Abroad *SS Muenchen, 72, 73*
 1930 — North River terminal, *72*
 1942 — *Normandie,* 83, 84, *84*
 1943 — Explosion on *El Estero, 84,* 85
 1945 — Boiler room in *El Estero,* 84
 1946 — St. George Ferry Terminal, *69, 70*
 1947 — Grace Lines Terminal, *91*
 1953 — Engine room explosion, *98*
 1962 — Sinclair Refining Co., warehouse, *125*
 1973 — Collision and fire, 71
 1973 — *Sea Witch* and *Esso Brussels,* 108, *108,* 109

New York harbor, 71

New York Naval Shipyard, 71

New York Shipbuilding, *27*

*New Yorker,* 9, *10,* 12, *12,* 16, 17, 18, 24, *24,* 26, *27, 44*

Newell, Harry, *117*

*Newell, Harry, 117*

Nichols Brothers Boat Builders, Inc., 117

N.C. Nickum & Sons Company, Inc., 56, *56,* 116

Nickum & Spaulding Associates, Inc., *115,* 116, 117, *118*

Nimitz, Chester W., 78

Norfolk Shipbuilding & Drydock Company, *100*

*Normandie,* 83, 84

North German Lloyd Company, 24, *25*

Norton Point in Gravesend Bay, 109

*Nortrans Visions, 111*

Norway boat
 *Gisla,* 63

Norwegian freighter
 *Gisla,* 63

Nott steam fire engine, *20*

*Novelty* pumper, 17

Oakland and San Francisco Firefighters, 80

Oakland Army Base, Outer Harbor, 79

Oakland boats
 *City of Oakland (Port of Oakland) (Hoga),* 78, 79, *80,* 100,
  *117*

Oakland fires
 1948 — *Hawaiian Rancher,* 79
 1949 — Oakland Army Base, 79
 1986 — Oil tanker explosion, *117*

Ocean liners
 *America,* 64
 *Kaiser Wilhelm der Gross,* 24
 *North German Lloyd,* 24, *25*
 *Saale,* 24
 *United States,* 64

*Ocean Steelhead,* 124

*Oglala,* 76

Oil refineries, 34
 N.K. Fairbanks and Standard oil refineries, 34

"Old Ironsides", *116*

Omega gearing, 118

Opera Comedy, 58
 Gilbert and Sullivan, 58

*Orient Trader, 102*

*Osborn, Lewis,* 20

Pacific Coast Engineering Company, 78

Pacific Coast Shipbuilders, 38

Pacific Drydock and Repair, 78

Pacific fleet, 78

Pacific Headquarters Base, USN, 75

Pacific Northwest, 123

Cities
    Seattle, 115
    Tacoma, 115
    San Francisco, 118
    Los Angeles, 118

Panama freighter
    *El Estero*, 84, *84*, 85, *85*, 99

Panama Canal, 39, 41, 45

Parmelee, Henry, 20

Parsons, Harry deBerkeley, 9, 12, *13*, 16, 18, 21, *25*, 27

Paterson, John L., 108
    Master of *Sea Witch*
    Died of heart attack in 1973, 108

Pearl Harbor boat
    *Hoga (City of Oakland) (Port of Oakland)*, 74, *74*, 75, 76, 77, *77*, 78, *78*, *117*

Pearl Harbor fires
    1941 — Pearl Harbor attack, *74*
    1941 — *Battleship Row* explosions, 75
    1941 — USS *Arizona* explosion, 75
    1941 — Ships torpedoed and explosions, 75
    1941 — *Nevada* casualties, 75
    1941 — *USS Oklahoma* sank, 75
    1941 — Bombings, 76
    1941 — Cruisers and Destroyers bombed, 76
    1941 — *USS Nevada* crippled, 76, *77*, 78
    1941 — Destroyer *USS Shaw* bombed, 76
    1941 — *Arizona* flames, 76, 78

Pearl Harbor
    Headquarters for United States Pacific fleet, 74, *74*, 75, 76, 77, 78, *78*, 79, *79*, 80

Petrich, James, 124

Petty officer in Japan
    Kanai, Noburo, 75, 78
    Also called Bombardier Kanai, 78

Philadelphia boats
    *Blankenburg, Rudolph*, 94
    *Delaware*, *93*, 94
    *Franklin, Benjamin (Franklin)*, *93*, 94
    *Moore, J. Hampton*, 94
    *Samuel, Bernard*, *92*, 93, *93*
    *Stuart, Edwin S.*, *23*, 94

Philadelphia fires
    1955 — Publicker Alcohol Company, *93*
    1961 — Abroad freighter *Polanic*, *94*
    1974 — Railroad yards, *111*

*Phoenix, 80*, 94, *95*, 97,. *97*, 98, 103

Phoenix Society fire buff club
    Prendergast, Edward, 97

Piano factory owner
    Parmelee, Henry, 20

Pier A., Battery Park, *111*
    Fireboat station, *111*

Pier 48, San Francisco, *86*

Pilots
    Dahlquist, Bill, 2, *2*, 8
    Fitzsimmons, Matthew T., Jr., 109
    Fridell, William, *62*
    Gray, Brainard "Choppy", 85, 90, 91
    Harvey, John J., *72*
    Hylant, Thomas, 58
    Nelson, Jim, *120*

Pipeman
    Hanson, Hans, 34

Planagan, John, 90, 91

Plant Shipyards Corporation, 97

Point DeFiance Boathouse, 124

*Polanic, 94*

Port Chicago, Ca., fire
    1944 — Ammunition ships explosions, *84*

Port Colborne, Ontario, fire
    1960 — Maple Leaf Milling Company, *60*

*Port Houston*, 45

*Porto Rican, 117*

Portland, Me. boats
    *City of Portland, 100*
    *Engine 7, 37*

Portland, Oregon, boat
    *William, George H.*, 29, *29*

Ports
    Port Chicago, Ca., *84*
    Ports of Great Lakes, 9, 11, 19
    Port of Los Angeles, 2, 52, 105, 109, 114
    Port of Los Angeles, Supt. terminal, 109, *112*, 113
    Port of New Orleans Dock, *45*
    Port of New York, 24, 84, 108, *108*
    Port of Oakland, Ca., 78, *79*
    Port of Seattle, 115
    Port of Tacoma, 122
    Port of Pacific Coast, 29

*Potter, George R.*, 9, *11*, 58, 59, 65

Prendergast, Edward, 97

Prohibition years, 1919-1933, 58
    Repealed, 58
        21st Amendment to U.S. Constitution, 58

Project Manager

Gow, Paul A., 116, 118

Public Works System, 30

Publicker Alcohol Company, *93*

Puget Sound Navy Yard, 77

Quartermasters
Brown, Bob, 74, 75, 77, 78
Johnston, Roy G., 76

Quayle, Frank, *91*

Queen, Edgar M., 89

Ramaekers, Gerald, 2

Ramsey, Sir Bertram, 82

Rasmussen, John, 2, 14

Ratcliffe, Dr. Allen W., 62

Ray, Henry Albert William, 82

Reddan, Danny, *66*

Reid, Daniel G., *111*

Reiser, James W., 122

Repair Ship
*USS Vestal,* 75

Richmond Beach Shipbuilding Co., 39

*Ring, Thomas,* 55

Rivers
Buffalo River, 58
Calumet River, *62, 64*
Chicago River, 19, *83, 95,* 97
Delaware River, 92, *94*
East River, 9, 17, 24
Mississippi River, 17, 45
Niagara River, 58
North River, 24, *72, 73,* 83
Thames River, *15,* 16, 17, 80, 82
St. Lawrence River, 35

Robbins Reef Lighthouse, *84,* 85

Robert E. Derecktoy's Shipyard, *69*

Rock Island Railroad Grain elevators, *95*

*Roen, John III (Illinois),* 34

Rolls-Royce of fireboats, 62

Roquemore, Joseph V. "Rocky", *84,* 85

Rotary Lift Company, 98

Royal Naval officer, navigator, 80

Royal Navy and Royal Marines, 82

RTC Shipbuilding Corporation, 92, *96*

*Saale,* 24

Sabotage fear, 83

Saint Lawrence Seaway, 91, *91*

*Samuel, Bernard, 92, 93,* 93, 94

J. Samuel White Shipyard, 80

San Diego boat
*Kettner, Bill, 42,* 43

San Francisco Bay, 98

San Francisco boats
*Governor Irwin, 30*
*Governor Markham, 30*
*Phoenix, 80,* 94, *95,* 97, *97,* 98, 103
*Scannell, David,* 30, *31, 32, 33,* 97
*Sullivan, Dennis T.,* 30, *31, 32, 33,* (&

San Francisco fires
1906 — Earthquake and fire, 30, *30, 31*
1972 — The Embarcadero, *110*

San Francisco-Oakland Bridge, 79

San Francisco Public Works System, *31*

San Pedro fires
1926 — Lumber schooner burned, *50*
1945 — Los Angeles Shipbuilding and Drydock Corp., *86*
1954 — Marine Terminal storage tanks, *97*

*Sansinena,* 109, *112,* 113, *113,* 114

Sarin, Jack, *117*

*Scannell, David,* 30, *31, 32,* 33, 97

Schickenbantz, Henry, 58

*Schlaeger, Victor, L., 94,* 97

Schwab, Frank X., 58, 59

*Scott, Ralph J. (Fireboat No. 2),* 2, 4, 6, 8, 45, *47, 48,* 51, *52, 54,* 56, *56,* 59, 67, 106, *107, 111, 113,* 114

Scott, Ralph J., *46,* 49, 51

Scott, Mrs. Ralph J. (Adeline), 49, 52

SCUBA, Boston firemen, 104

SCUBA diver
Hilzer, Charles C., 106

SCUBA Firefighters, 6, 8, *8, 102,* 103, 104, 106
Divers, 3, 4, 6, 8, *101, 102, 104, 106,* 106, *112*
Fire fighting program, 105
Instructors, 103
LAFD firefighters, 105
L.A. SCUBA team, 105
LAFD skin divers, 105

SCUBA operations, 121

*Sea Witch,* 108, 109, *109*

Seaplane tender boat
   *Avocet,* 176

Seattle boats
   *Admiral Farragut,* 40
   *Athlon,* 40
   *Alki,* 56, *56,* 58, 59, 115, *115,* 116
   *Chief Seattle,* 115, *115,* 117, 118, 119
   *Duwamish,* 38, *38,* 39, *39,* 40, *40,* 41, *41, 56, 115*
   *Snoqualmie,* 20, *21,* 40, 41

Seattle fires
   1899 — Waterfront and commercial district, 20, *21*
   1914 — Grand Trunk Complex, *38,* 40, 42
   1984 — West Marginal Way, 41

Shand, Mason & Co., *14,* 15

Shaw, Sir Massey Eyre, 80, *81*

Shell Oil Terminal, *90, 90*

Siddal Fisheries, Ontario, 59

*Sierra, 50*

Silsby Manufacturing Co., 9
   Steam engine builder, 9

*Silver Ash,* 71

Sinclair Refining Co., *125*

Skin diver
   Beatteay, Bob, 105
   See Firemen

Skin diving (Buddy System), 6, 103

SKUM blabber mouth-type nozzle, *118*

Sinclair Refining Company warehouse, *125*

Sing Sing prison, 26
   Van Schaick, Captain William H.,
   prisoner at Sing Sing, 26

Smith, "Buzz", 4

Smith, John E. *100*

*Smoke* II, 98

Smith and Watson Iron Works, 53

*Snoqualmie,* 20, *21,* 40, 41

Society of Naval Architects and Marine Engineers, 9, 28

Socony (Mobil) Oil Dock, 58

South Michigan Avenue Bridge, *60*

South quay waterside development, 83

*SS Muenchen,* 72

Sperry electric-and hand steering combination, 93

Standard Shipbuilding, *44*
   Shooters Island, New York

Stang nozzles, 105

Stang Hydronics Master monitor, 124

Stang Intelligiant bow monitor, *111, 116*

Stang Shaper monitor, 118

Staten Island, 108

Statue of Liberty, *65,* 84

Steamship
   *General Slocum, 24,* 25, 26
   Mississippi River Steamboat, 17
      *Main,* 24, 25

Sterling-Viking Engine, *62, 87*

Stetson, Frank L., 40

Stevens, Arthur D., 43, 44, 45

*Stewart, Graeme,* 34, *36,* 44

Stockton, Ca. boat, *87*
   *Hawaiian Rancher,* 79, *79,* 100

Stoddard, Fred, 3, 6

Stokes, John W., 40

*St. Florian, 112*

St. Mark's Lutheran Church, 26
   *General Slocum* fire enroute to picnic, 26

Strebor gun, 124

Strong, Arthur, *104*
   Battalion chief, *104*
   Died in line of duty in 1963, *104*

*Strong, William L.,* 24, 26

*Stuart, Edwin S., 23,* 94

Sturgeon Bay Shipbuilding & Drydock Company, 59

Submarine-Chaser, *S.C. 145,* 43, 44
   Changed to fireboat *Callahan, John H.,* 43, 44

Submarine-Chaser, *S.C. 428, 43*

*Sullivan, Dennis T.,* 30, *31, 32, 33, 86,* 97

*"Super Fireboat",* 12

Super Pumper System, *102*
   Satellite hose, *102*
   Super pumper, *102*
   Super tender, *102*

Superintendent of London Fire Engine Establishment, *16,* 17
   Braidwood, James, *16,* 17

Surface-effect-system SES, 119, 120, *120*
   Multi-purpose fireboats, 105